Heeding the Call for Change

Suggestions for Curricular Action

Lynn Arthur Steen, Editor

THE MATHEMATICAL ASSOCIATION OF AMERICA

MAA Notes and Reports Series

The MAA Notes and Reports Series, started in 1982, addresses a broad range of topics and themes of interest to all who are involved with undergraduate mathematics. The volumes in this series are readable, informative, and useful, and help the mathematical community keep up with developments of importance to mathematics.

MAA Notes

1. Problem Solving in the Mathematics Curriculum,
   Committee on the Teaching of Undergraduate Mathematics,
   a subcommittee of the Committee on the Undergraduate Program in Mathematics, *Alan H. Schoenfeld,* Editor.

2. Recommendations on the Mathematical Preparation of Teachers,
   Committee on the Undergraduate Program in Mathematics, Panel on Teacher Training.

3. Undergraduate Mathematics Education in the People's Republic of China,
   Lynn A. Steen, Editor.

4. Notes on Primality Testing and Factoring,
   Carl Pomerance.

5. American Perspectives on the Fifth International Congress on Mathematical Education,
   Warren Page, Editor.

6. Toward a Lean and Lively Calculus,
   Ronald G. Douglas, Editor.

7. Undergraduate Programs in the Mathematical and Computer Sciences: 1985–86,
   D. J. Albers, R. D. Anderson, D. O. Loftsgaarden, Editors.

8. Calculus for a New Century,
   Lynn A. Steen, Editor.

9. Computers and Mathematics: The Use of Computers in Undergraduate Instruction,
   Committee on Computers in Mathematics Education, D. A. Smith, G. J. Porter, L. C. Leinbach, and R. H. Wenger, Editors.

10. Guidelines for the Continuing Mathematical Education of Teachers,
    Committee on the Mathematical Education of Teachers.

11. Keys to Improved Instruction by Teaching Assistants and Part-Time Instructors,
    Committee on Teaching Assistants and Part-Time Instructors, Bettye Anne Case, Editor.

12. The Use of Calculators in the Standardized Testing of Mathematics,
    John Kenelly, Editor, published jointly with The College Board.

13. Reshaping College Mathematics,
    Committee on the Undergraduate Program in Mathematics, Lynn A. Steen, Editor.

14. Mathematical Writing,
    by *Donald E. Knuth, Tracy Larrabee, and Paul M. Roberts.*

15. Discrete Mathematics in the First Two Years,
    Anthony Ralston, Editor.

16. Using Writing to Teach Mathematics,
    Andrew Sterrett, Editor.

17. Priming the Calculus Pump: Innovations and Resources,
    Committee on Calculus Reform and the First Two Years,
    a subcommittee of the Committee on the Undergraduate Program in Mathematics, *Thomas W. Tucker,* Editor.

MAA Reports

First Printing
© 1992 by the Mathematical Association of America
ISBN 0-88385-079-6
Library of Congress Card Catalog Number 92-060332
Printed in the United States of America

Heeding the Call for Change

Suggestions for Curricular Action

Preface

The "call for change" issued by the Board of Governors of the Mathematical Association of America (in *A Call For Change*, 1991) may appear at first glance to be only about the mathematical preparation of teachers, as its subtitle proclaims. But two ingredients combine to give it much broader scope. First, the logic of change is rooted in insight into how all students learn, and is not limited to those few who are preparing to be teachers of mathematics. Second, the call extends to the entire undergraduate curriculum since most students preparing to be secondary school teachers take mathematics courses alongside all other students. Thus *A Call for Change* heralds sweeping reform in all aspects of collegiate mathematics.

Heeding the Call for Change provides the first in a series of challenges concerning where and how to begin the process of change. The themes covered in this volume are quite diverse, ranging from disciplinary discussions (e.g., statistics, geometry) to curricular systems (e.g., the undergraduate major), from administrative concerns (e.g., assessment) to policy debates (e.g., multiculturalism). Yet beneath the surface of these varied papers lie many of the fundamental themes found in *A Call for Change*: that instruction needs to become an active, constructive process in which students learn to communicate about mathematics, to build mathematical models, and to connect mathematical ideas with the world around them.

Chapters in this volume are diverse not only in subject, but also in source. Five chapters (on statistics, quantitative literacy, geometry, environmental mathematics, and assessment) are the products of electronic e-mail Focus Groups. These groups were established by the MAA as a strategy for generating and recording informed debate on topics of current interest. Each group operated for a period of approximately two or three months under the leadership of a moderator, who subsequently prepared a report from the group based on the record of e-mail discussion. Many of the chapters contain appendices that help provide a more complete record of the issues under discussion.

Two other chapters are more journalistic in spirit, each being a report based on telephone interviews and written documents in two areas of lively debate—multiculturalism and educational research in collegiate mathematics. The final two chapters are reprints of important MAA reports that have appeared previously in other more ephemeral sources: the 1990 report *Challenges for College Mathematics* of MAA and the Association of American Colleges (AAC), and the 1991 report of CUPM on the undergraduate major.

The opening chapter, *Teaching Statistics*, brings to the surface the need for change in approach to instruction—less lecture, more exploration—and in emphasis of curriculum—more data, less theory. Moderator George Cobb is Dean of Studies and Professor of Statistics at Mount Holyoke College and chair of the joint Committee on Statistics of the MAA and the American Statistical Association (ASA). A related report from a 1990 workshop organized by Robert V. Hogg is reprinted in an Appendix.

Tomorrow's Geometry, moderated by Joe Malkevitch of York College, reveals the enormous gap between the rich intellectual resource of contemporary geometry and the barren soil of typical college geometry courses that are aimed at preparing students to teach traditional high school geometry. An Appendix to this chapter provides a comprehensive outline of possible new curricula for college geometry courses.

Environmental Mathematics, the third in the discipline triad, introduces an entirely new field of emphasis for undergraduate courses in mathematical modelling, a field of immense interest to students and of great importance to us all. Moderator Ben Fusaro of Salisbury State University is chair of the MAA Committee on Environmental Mathematics. A report from *The Chronicle of Higher Education* on environmental mathematics that was presented at the January 1992 AMS/MAA meeting in Baltimore is reprinted in an Appendix.

Reaching for Quantitative Literacy surveys the many approaches and arguments that surround the vexing question of what mathematics all college graduates should be expected to know. This issue takes on greater importance and difficulty as new standards work their way into school mathematics. As is clear from this Focus Group discussion, it is difficult to frame college expectations in a form that both extends the new school mathematics standards and also is realistically achievable in colleges. Moderator Linda Sons of Northern Illinois University chairs the CUPM Subcommittee on Quantitative Literacy.

In *Multiculturalism in Mathematics: Historical Truth or Political Correctness?*, Allyn Jackson surveys the implications for mathematics (and the attendant controversies) of the efforts to teach mathematics with a more international and less "Western" flavor. Advocates and critics debate the merits of relating the mathematics curriculum to students' cultural or ethnic origins. The issue of multiculturalism, already of great urgency in the humanities, may soon be more widely debated in mathematics departments as well.

Assessment of Undergraduate Mathematics moves even further into the policy arena, examining the goals and objectives of program reviews that are suddenly becoming *de rigueur* in colleges and universities across the nation. The importance of measuring what is of value rather than what is available becomes clear as one reads the comments of these Focus Group participants, including both mathematicians and administrators with considerable wisdom and expertise in program assessment. Moderator Bernard Madison is Dean of the Fulbright College of Arts and Sciences at the University of Arkansas, Fayetteville, and chair of the CUPM Subcommittee on Assessment. An Appendix reprints a seminal article on assessment by Grant Wiggins which every mathematician and department chair should read.

Untying the Mind's Knot, similarly, is about matters of controversy—about whether educational research that seeks to understand how college students learn mathematics leads to fundamental principles that stand the tests of objectivity and replicability normally associated with traditional research. Reporter Barry Cipra reflects in his analysis on the various schools of thought concerning the value of educational research on undergraduate mathematics, and on the likelihood that results of this research can actually improve instructional practice. An Appendix contains the report of an MAA-sponsored conference on this subject held in November 1991, including several recommendations for action by the MAA and other professional societies.

The final two chapters provide contemporary recommendations on the content and context of the mathematics major. The first, *Challenges for College Mathematics: An Agenda for the Next Decade*, is the 1990 report of the MAA-AAC Task Force on Study in Depth that discusses issues of departmental climate, community, and culture that are ingredients of successful undergraduate programs in mathematics. The second, *The Undergraduate Major in the Mathematical Sciences*, is a 1991 report of CUPM that gives detailed advice about curriculum, tracks, and assumptions that should undergird any sound mathematics major.

Each chapter in this volume highlights many options for constructive change; most also offer specific suggestions for improvement in curriculum or instructional practice. They provide not a blueprint but a general framework within which much needed improvement in undergraduate mathematics can take place. Departments that begin to explore the ideas found in the chapters of this volume will indeed be heeding the call for change.

Acknowledgements. Support for the preparation of this volume was provided by the National Science Foundation through a grant to the MAA entitled "Curriculum Action Project." The intent of this project is to stimulate the college and university mathematical community to undertake important improvements across the entire spectrum of undergraduate mathematics. Oversight for the project was provided by a National Steering Committee, whose members are listed on the following page. On behalf of the Committee, I would like to thank all those who contributed to this volume—members of the Focus Groups, various reviewers, writers, and others who generously contributed course outlines, bibliographies, or special reports.

Lynn Arthur Steen
St. Olaf College
January, 1992

Curriculum Action Project
National Steering Committee

Teaching
Statistics

Teaching Statistics

George Cobb

MOUNT HOLYOKE COLLEGE

Introduction

This report on teaching statistics will present the Statistics Focus Group's recommendations under three headings, corresponding to *statistics* ("Recent Changes in the Field"), *mathematics* ("Some Differences Between Mathematics and Statistics"), and *teaching* ("What Research Tells Us"). A fourth section ("Examples") illustrates ways these recommendations can be put into practice, and a final section ("Making It Happen") offers two meta-recommendations about implementation.

> What do we want students to be able to *do*, themselves, in terms of performing statistical work after their course is completed? What kinds of statistical reasoning, or arguments, do we want them to be able to *understand*? What kinds of *experiences* should the students have had in the course?
>
> *—Jim Landwehr, AT&T Bell Labs*

Our group made a deliberate decision not to prescribe lists of topics for particular courses, but instead to seek a general intellectual framework within which we and others can fit a great variety of courses. (For one list of recommended topics, see Appendix D: Report on a Workshop on Statistical Education, by Robert Hogg, which presents a consensus view of thirty-nine statisticians on problems in statistical education and suggestions for reform.) We intend our three recommendations to apply quite broadly.

> One introductory course may differ from another according to many factors: calculus prerequisite versus no calculus; engineering, technical audience versus arts, nontechnical audience; goal of understanding versus goal of doing; taught by mathematics or statistics department versus taught by user department; large research university versus small college; large clientele (100s–1000s) versus small clientele (less than 100); required course versus elective course; students bright, intellectually curious versus students dull, passive; PC's readily available versus computer facilities inadequate.
>
> *—Howard Taylor, University of Delaware*

> If I use applied regression as the vehicle to give students the experience they need and you use time series forecasting, that's fine. What matters most is the experience with practical reasoning about data.
>
> *—David Moore, Purdue*

Although our group spent most of its time thinking about several versions of the "standard" introductory course, several of us know from experience that the spirit of what we urge can infuse courses devoted to experimental design and analysis of variance, sample survey design, regression, time series, or multivariate analysis, to name just five. At Swarthmore, Gudmund Iversen teaches a pair of introductory courses ("Statistical Methods" and "Statistical Thinking") which are in some ways quite dissimilar, but the kind of general goals we shall describe are appropriate for both. At Mount Holyoke, the mathematics department offers two beginning statistics courses, one on experimental design and applied analysis of variance, with emphasis on experimental studies in biology, psychology, and medicine, and another on applied regression, with emphasis on observational studies in public policy, economics, etc. At St. Lawrence University, Robin Lock teaches an introductory course on applied time series. At University of Delaware, Howard Taylor teaches beginning statistics to engineers. At SUNY Stony Brook, Judy Tanur teaches statistics to sociologists, and does a lot with surveys. The goals we advocate are suitable for all these courses.

We have also sought to make our three recommendations relatively independent of level of preparation, in that we imagine them applying to courses designed for students at any band within an entire spectrum of technical facility, ranging anywhere from ultra-nimble to infra-numb. Level of preparation will determine how far or how deep a course can go, but need not determine the general direction; it may limit the degree of subtlety and sophistication with which students can come to understand random variability and it's underlying predictable patterns, but needn't prevent them from making a meaningful start.

Recent Changes in the Field of Statistics

> Statistics has moved somewhat away from mathematics back toward its roots in scientific inference and the analysis of data. ...The most important driving force in this shift of emphasis is the computer revolution.
>
> —David Moore

During the last two decades, statistics has been changing simultaneously on three levels, which correspond to technique, practice, and theory. On the technical level, cheap, powerful computing has made possible a number of important innovations: graphical methods for data display, iterative methods for data description, diagnostic tools for assessment of fit between data and model, and new methods of inference based on resampling techniques such as the bootstrap. On the level of practice, such things as pattern-searching, model-free description, and systematic assessment of fit have all become more prominent, at the expense of formal inference, most especially hypothesis testing. Statisticians now put more effort into the complex process of choosing suitable models, less effort into doing those things—simpler by comparison—which take the choice of model as given.

> How many of us still concentrate on hypothesis testing, and even within that narrow and questionable context permit students to finish an exercise in number crunching to a pre-specified t-test, with the sole application of the English language being the words "reject the null hypothesis?"
>
> —Walt Pirie, VPI & SU

> Mathematicians who teach the introductory course probably will be completely oblivious to the decreasing importance of hypothesis testing in the work of statisticians. To mathematicians, this may have the most profound implications for their introductory course because it calls into question the ultimate goals of the course.
>
> —Ann Watkins, California State at Northridge

On the level of theory, one can distinguish two kinds of changes that invigorate discussions about the reasoning of statistics. First, foundational discussions of long standing (should unknown parameters be regarded as fixed or as values of random variables? Should inferences be conditional or unconditional?) now much more often take place in the context of real applications. Second, statistical practice has partly outgrown its mathematical theories, which are consequently less relevant. Important new elements of data analysis (model-choosing, model-checking, and model-free description) don't fit the older theoretical frames, while the influential area of statistical process control offers new ways, not yet mathematically developed, to frame the enterprise of learning from data.

> [Alternatively,] a coherent theme built around modern information processing can encompass the general concepts of EDA (exploratory data analysis), graphics, and computing as they relate to the introductory course in statistics.
>
> —*Dick Gunst, Southern Methodist*

A deeper and more detailed treatment of the recent changes in statistics may be found in David C. Hoaglin and David S. Moore (Eds.), *Perspectives in Contemporary Statistics*, MAA, 1992. Suggestions for additional reading can be found at the end of this report.

What does all this mean for the teaching of statistics? We offer the following recommendation:

Recommendation I: EMPHASIZE STATISTICAL THINKING

Any introductory course should take as its main goal helping students to learn the basic elements of statistical thinking; many advanced courses would be improved by a more explicit emphasis on those same basic elements:

1. *The need for data.*

> To recognize in one's own citizenship, the need to base personal decisions and actions on evidence (data), and the dangers inherent in acting on assumptions not supported by evidence. This doesn't even necessarily invoke the concept of randomness, but is nevertheless inherent in statistics as a discipline.
>
> —*Walt Pirie*

2. *The importance of data production.*

> It is very difficult and time-consuming to formulate problems and to get data that are of good quality and really deal with the right questions. Data generally don't represent what people initially think. Moreover, most people don't seem to realize that this is the way things work out until they go through this experience themselves. This is the most important part of actually doing statistics, because if it is not done well all the subsequent analysis can't be worth much. ...Most people I deal with would be better off if they carried realistic notions about formulating problems and getting relevant and accurate data, rather than some vague notion of significance or confidence from the course they had.
>
> —*Jim Landwehr*

> I haven't had students plan studies and gather data in a couple of years, for various reasons. I wasn't very happy with the actual projects during the four semesters I did it. However, I was quite happy with the students' *experiences*. Almost every one of them, by the time they finished, was rather sheepish about what a poor study it turned out to be, because they could see all the ways it really should be improved.
>
> —*Mary Parker, Austin CC*

> Let's not forget that the most desirable process is usually (always?) starting with the model, in the sense of "design before data." In my experience the most frequent cause of poor (or even failed) experimentation is the lack of good prior design.
>
> —*Walt Pirie*

3. *The omnipresence of variability.*

> Variability is ubiquitous. It is the essence of statistics as a discipline . . . it is not best understood by lecture—it must be experienced.
>
> —*Dick Gunst*

> Some data have variability due to measurement error, where other data have variability due to the fact that the phenomenon isn't completely deterministic. As mathematicians, we notice that the same models can be used to analyze these two different kinds of variability, so we tend to identify them with each other. I think that students sense that these are different, and this is part of why they have a problem with the idea of a random variable. I had my students measure the length of the building by stepping it off, to illustrate that many measures are approximate even if we report them as exact. The scores for the whole class formed a normal distribution, with a few outliers, and we could see why we needed to discard them. In the same assignment, I asked each student to measure the heights of three females. These scores for the whole class also formed a normal distribution. We could see why it wasn't appropriate to discard outliers this time.
>
> —*Mary Parker*

4. *The quantification and explanation of variability.*

   a. Randomness and distributions.
   b. Patterns and deviations (fit and residual).
   c. Mathematical models for patterns.
   d. Model-data dialogue (diagnostics).

> —*David Moore*

> The development of the *history* of statistical thinking should also be included in any general education statistics course. (And not in a separate unit, but infused throughout.) Part of the reason that many instructors resist changes in the introductory statistics course is that they don't understand how relatively young the subject is and that evolution can be expected.
>
> —*Ann Watkins*

> Just as there was mathematics before Euclid and Archimedes, there was statistics before Karl Pearson and Ronald Fisher, but comparing the times when these four lived gives a sense of how new a subject statistics is. In the context of two millenia of changes within mathematics, essentially *all* changes in statistics are recent changes.
>
> —*George Cobb, Mount Holyoke*

Some Pertinent Differences Between Statistics and Mathematics

Much thinking about statistics falls victim to an unfortunate habit of locating quantitative courses along an intellectual continuum which stretches from the mechanical to the mathematical, from recipes (bad) to theory (good). Two natural but destructive consequences of this one-dimensional view are that data tend to get associated with recipes and dismissed as devoid of intellectual challenge, and mathematical theory is taken to include all of statistical thinking, so that non-mathematical concepts of statistics are not taught. Real life is not so simple.

> Statistics in practice involves a dialogue between models and data that is quite different from the deductive derivation of properties of models met in theory-based courses. Models are used to analyze data, but the data are allowed to criticize and even falsify a proposed model (diagnostics). Several different methods may be applied fruitfully to the same problem. The results of one study

often suggest another study, not a final conclusion. Students need to meet and use this way of thinking.

—David Moore

The key is to teach statistics like statisticians instead of like mathematicians.

—Walt Pirie

Qualification: Math remains essential; you can never be too rich or too thin or know too much mathematics.

—David Moore

The question is, who is to be master—that's all.

—Humpty Dumpty, Through the Looking Glass.

Recommendation II: MORE DATA AND CONCEPTS: LESS THEORY, FEWER RECIPES

Almost any course in statistics can be improved by more emphasis on data and concepts, at the expense of less theory and fewer recipes. (To the maximum extent feasible, automate calculations and graphics.)

—David Moore

Statistical concepts are best learned in the context of real data sets. Fortunately, using the computer to automate calculations and graphics makes it possible to work with real data without becoming a slave to the mechanics.

The introductory course in statistics should focus on a few broad concepts and principles, not a series of techniques. Suggested concepts are: graphing data (as in Cleveland's book—this is not trivial), randomness (the idea of producing a predictable pattern through randomness is difficult for a student to grasp; it is not intuitive), inferential reasoning (ideas illustrated through bootstrap-like simulations are easiest to grasp and the formulas, for those who insist on using them, are approximations), experimental design (I've seen eyes light up for some reasonably intelligent people when they were able to set a stat-a-pult right on target after collecting information on a half rep of a two to the fourth factorial experiment; they did not think it could be done).

—Dick Scheaffer, Florida

I am less concerned with the mechanics than I am with students being able to display data quickly by hand or computer and then *interpret* the display. Examine many of the introductory texts today. See how many ask in chapter exercises for students to interpret the salient features of the graphs. The inability of students to interpret the graphics they produce makes the process merely an abstract exercise. So too with descriptive statistics. They are initially presented as a calculation exercise, without regard to their informational content.

—Dick Gunst

[...] was the first introductory textbook attempt that I know of to approach a data set with a set of candidate tools (parametric, non-parametric, and robust), and the attitude of trying to determine which would be best for this particular problem. When I embraced that approach, a lot of students were really enthusiastic about it. But the resistance by many of my colleagues was intense and intractable. Now the second edition is out, and sure enough, the non-parametric and robust material has been relegated to chapters at the end of the book like all other texts, where instructors can (and do) ignore it. Want to compare means of two groups? Push a button and out comes a *t*-test, the only thing we have to offer. No need to *think* about alternative approaches or make a judgment, just memorize the formula.

—Walt Pirie

Thinking about the role of probability can be a useful exercise in clarifying differences between statistical and mathematical thinking. Many statistical concepts don't rely on probability theory at all, and a course which puts statistical concepts ahead of mathematical theory will recognize that fact.

"Yes, Virginia, there is statistics without probability." Many basic ideas can be discussed prior to any discussion of probability. Most of EDA (exploratory data analysis) and graphing falls into this category, as does the basic principle of model building and experimental design. EDA allows for exploration and summarization of data and the formulation of questions, some of which may have obvious answers just from the exploration. Modelling looks at relationships but, at the basic level, does not require a goodness-of-fit test. (Does the scatterplot go up or down? Are the points close to a line or is there lots of scatter? Does interest in sports seem to be associated with the sex of the respondent?) Experimental design involves planned investigations to answer specific questions, but the plan and the data that result can be looked at without probability notions. (It seems that I can get to school faster by route A than by route B. Stereo A is really the better buy when you consider its reliability.

—*Dick Scheaffer*

Some important statistical concepts do, of course, depend on probability, but it is all too easy to ignore Humpty Dumpty and forget which is to be master.

Probability should be introduced in a less threatening way than we have traditionally done, and only in the framework of enabling students to draw statistical conclusions that enhance their simple graphs, simple statistics. The topics chosen should be directed toward a continuity or constancy of purpose (to borrow from Deming). I would like students to be able to use probabilities to make statistical conclusions: to understand the difference between phenomena that are likely and those that are unlikely, and, above all, to understand the distinction between phenomena that are "real" (in a statistical sense) and those that are likely to have occurred by chance.

—*Dick Gunst*

The distinction between mathematical theory and statistical concepts remains an important one even in thinking about the standard introduction to mathematical statistics.

I don't think students who take the standard mathematical statistics course come away with even the faintest appreciation for what statistics is about. Unless students have had a previous course that does justice to data analysis, and so provides a meaningful context for the mathematical statistics, the course is mainly an opportunity to practice advanced calculus techniques. I think only three positions are tenable here:

1. The mathematical statistics course should *never* be taught to students who haven't first taken an applied course;
2. The mathematical statistics course must be radically revised, to integrate data analysis with the statistical theory; or
3. The mathematical theory of statistics should be introduced via an optional adjunct to the beginning applied course.

—*George Cobb*

Those students who have had my Statistical Thinking course early (freshmen, sophomores) are having a ball with mathematical statistics as juniors or seniors. Others struggle more because they get bogged down by probability theory and mathematical niceties like moment generating functions, and they have a harder time seeing what statistics is all about. This points to a need to hear statistics twice before it makes sense, and we cannot lose the connection to real data.

—*Gudmund Iversen, Swarthmore*

My personal peeve about statistics courses in the typical undergraduate mathematics major, the upper-division offerings, is that they tend to 75% or more probability theory with possibly a second semester of mathematical statistics that most students do not take. Our mathematics majors often see little of the material taught in the introductory non-mathematical statistics courses for, say, psychology majors. This is especially true if a non-statistician teaches the course in the mathematics department. The 1981 CUPM report urged that the standard junior-level one-semester probability/statistics course taken by a mathematics major include a month of work with data. Most texts for this course still have nothing about working with data.

—Alan Tucker, SUNY Stony Brook

I think the probability–mathematical statistics sequence is important but should be preceded by a data analysis course.

—Jack Schuenemeyer, Delaware

In an ideal Grinnell I'd like my students to have an applied methods course as sophomores and then have a more traditional mathematical statistics sequence.

—Tom Moore, Grinnell

One of my colleagues was visiting from Norway, where they have a more extensive undergraduate statistics program than we do. Their experience was that many of the best students—those you particularly want to attract to statistics—were turned off by the early course without the mathematical underpinnings, and decided that they weren't interested in statistics. This surprised me at the time, but since then I have had a couple of quite bright mathematics majors take my elementary statistics course, with similar experiences. They seemed rather negative about statistics as we went through the elementary course, and neither was very interested in taking a mathematical statistics course because they felt they already knew all of that stuff. While I think I have convinced both of them to go on and take mathematical statistics, this experience indicates to me that we need a different beginning course than the typical elementary statistics course. And I suspect that the ideal beginning course for these mathematics people will not be the same as the ideal beginning course for the rest of our audience.

—Mary Parker

This sequence should not be the first course, or, at the least, should be accompanied by a lab that shows the other side (I almost said "the real nature") of the subject. For example, since the randomized comparative experiment is arguably the most important statistical idea of this century, one which has revolutionized the conduct of research in many fields of applied science, it's a sin to teach a first course that doesn't mention this idea and emphasize the contrast between observational and experimental studies. Because the two-sample t procedures and the mathematical model on which they are based ignore this distinction, the usual statistics theory course for mathematics majors also ignores it. That, as I said, is a sin. This example shows in brief what's wrong with the introduction we often give our majors.

—David Moore

Recent Research On How Students Learn

Shorn of all subtlety and led naked out of the protective fold of educational research literature, there comes a sheepish little fact: lectures don't work nearly as well as many of us would like to think. This rather discouraging assertion is supported by two clusters of research results. The first cluster shows part of what makes learning hard and lecturing often ineffective; the second shows the kinds of things that do seem to work when lecturing doesn't.

A. *Basic concepts are hard, misconceptions persistent.* As teachers, we consistently overestimate the amount of conceptual learning that goes on in our courses, and consistently under-estimate the extent to which misconceptions persist after the course is over.

> Ideas of probability and statistics are very difficult for students to learn, and conflict with many of their own beliefs and intuitions about chance and data. Students connect new ideas to what they already believe, and correct or abandon erroneous beliefs reluctantly, only when their old ideas don't work or are inefficient. Learning is enhanced by having students become aware of and confront their misconceptions.
>
> *—Joan Garfield, University of Minnesota*

> I am still chagrined by an experience in class several weeks ago. Using IQ scores $N(100, 15)$ generated by the computer, my class "discovered" the central limit theorem. No problems. They were fairly experienced with sampling distributions, expected the normal shape, knew that the expected value of the sample average equals the population mean, and weren't surprised by the standard error. We then went on to examine a case where the population was decidedly not normal, using the ages of pennies that we brought to class. Suddenly they had no idea where the expected value of the sample average should lie for their samples of size four. We spent a lot of time establishing that. I am chagrined because in previous semesters, using synthetic data, I had essentially assumed that the location of $\mu(\overline{x})$ was obvious to students, spending time instead establishing the shape and standard error of the distribution of sample means.
>
> *—Ann Watkins*

B. *Learning is constructive.* To absorb the full impact of these three words, you have to push their implied metaphor to its limits: concepts are constructs, learning is building. (Moreover, any building that students do will start with whatever conceptual raw materials they bring with them to the course. There's no such thing as starting from scratch.) By taking these construction images to the edge of the literal, and applying common-sense principles of carpentry to the process of teaching and learning, you can arrive at much the same conclusions as those who do research on how students learn: If you want to teach your students to build, you won't spend much time just talking, and what talking you do will occur on-site, where students who are learning-by-doing will want and need your comments on their work.

> Effective learning requires feedback. Students learn better by active involvement; they learn to do well only what they practice doing; they learn better if they have experience applying ideas in new situations.
>
> *—Joan Garfield*

Taken together, the two sets of results lead to a third recommendation:

Recommendation III: Foster Active Learning

As a rule, teachers of statistics should rely much less on lecturing, much more on the following alternatives:

1. *Group Problem Solving and Discussion.*

> I do not lecture at all. Instead, students are required to read the textbook, guided by a student handbook I have written containing study questions, sample problems, etc. Each day we first discuss the study questions, often arguing about issues. ... After our large group discussions, students then work in permanent small groups on activities, usually analyzing a set of data and discussing questions about these data sets.
>
> *—Joan Garfield*

2. *Lab Exercises.*

Statistics should be taught as a laboratory science, along the lines of physics and chemistry rather than traditional mathematics. Students must get their hands dirty with data. The laboratory must be a requirement and must contain more than just a few computers. This approach involves real data but also involves manipulative devices that include spinners, cards, bead boxes, a quincunx, stat-a-pults or similar devices for experimental design, and so on. Many things seem to work in the design of experiments realm (George Box's paper helicopters, Lego cars, rubber band guns, popcorn (see Hogg and Ledolter), balloons, melting ice cubes and on and on ...), the important idea is that an experiment is to be designed to answer a specific question and at least two important factors can be controlled.

—Dick Scheaffer

3. *Demonstrations Based on Class-Generated Data.*

Whether by counting the number of red M&M's in bags, taking surveys, or conducting simple experiments, the scientific enterprise referred to as the field of statistics must be experienced. This is very difficult to do in large lecture sections, so opportunities for demonstrations in lieu of hands-on individual experiences are a necessary alternative. Ideally, several demonstrations or experiences would occur in a single course. They will, I contend, prove to be one of the features of the course that will be remembered long after the formal analyses are forgotten.

—Dick Gunst

4. *Written and Oral Presentations.*

Students come to us with primarily an intuitive understanding of the world. It is part of our job to ferret out those intuitive processes and correct the incorrect ones. As far as I know, this can only happen by having students discuss and write about their understandings and interpretations of problems.

—Dick Scheaffer

5. *Projects, Either Group or Individual.*

Students in a first course should learn by doing. They will buy into the course and subject if they can formulate and design projects, collect and analyze data. Easy to use statistical software with graphics should be emphasized. Calculations should be de-emphasized.

—Jack Schuenemeyer

I'm doing student projects for the first time in an introductory course, and I have never had such enthusiasm. The assignment is to do a survey about some issue of interest at California State University at Northridge.

—Ann Watkins

Examples

Ah, generalities, which like fish glitter but stink. Here we escape glittering generalities.

—David Moore

Our proposed escape route is marked by two clusters of examples. The first cluster relates principally to our first two recommendations, on what to teach, and consists mainly of examples of entire courses. Taken together, these examples illustrate how remarkably different courses may be, both as to technical level and as to statistical topics, while nevertheless serving the general goals we have spelled out. The second cluster of examples relates principally to our third recommendation, on how to teach, and consists mainly of examples of parts of courses. These examples offer teachers of statistics a variety of alternatives to lecturing, all of them compatible with the examples of course content from the first cluster.

Teach Statistical Thinking

The following examples illustrate how the maxims "more data, fewer recipes" and "more concepts, less theory" improve students' statistical thinking. We begin with the junior high and high school level (Quantitative Literacy Project), then present three undergraduate general education courses (Chance, Quantitative Reasoning, and Statistical Thinking), then three courses that are technically somewhat more demanding (Statistical Methods, Time Series, and Multivariate Statistics), and end with two quite different variants of the standard mathematical statistics course.

These courses rely on computers in various ways. For example, both Mount Holyoke's quantitative reasoning course (No. 3) and Oberlin's mathematical statistical course (No. 7) use computers to analyze moderate-to-large archival data sets. (Most of the other courses also use computers to analyze data sets.) A very different approach to the mathematical statistics course (No. 8) relies on computers for simulation-based empirical investigation of the properties of estimators.

1. *The Quantitative Literacy Project.* (Dick Scheaffer, University of Florida)

The National Council of Teachers of Mathematics (NCTM) recently released their *Curriculum and Evaluation Standards,* which have a carefully delineated strand in statistics throughout the curriculum and an emphasis on modelling from data in other areas such as algebra and functions. That emphasis on data analysis should be woven through the mathematics curriculum and connected to other components of the curriculum is seen in the following:

> This standard should not be viewed as advocating, or even prescribing, a statistics course; rather, it describes topics that should be integrated with other mathematics topics and disciplines.

The NSF-funded Quantitative Literacy Project (QLP), a joint project of the American Statistical Association (ASA) and the National Council of Teachers of Mathematics (NCTM), served as the basis from which the strand in statistics was developed for the *Standards.* The QLP provides curriculum materials in certain areas of data exploration, probability, and inference, in a style that makes the material accessible to teachers and students, and provides a model framework for in-service programs to enhance the skills of teachers in the area of statistics and probability.

The curriculum units developed by the QLP explore elementary topics in data analysis, probability, simulation, and survey sampling, with new units being planned for exploring measurements and planning experiments. The approach is to use real data of interest to the students and simulations of real events to show how to use statistical ideas to extract useful information from numbers. Many of the statistical tools are graphical and reflect the latest thinking among practicing statisticians. Teachers using these materials are provided with opportunities to make heavy use of hands-on activities, group discussion, student projects, and report writing. (From "The ASA-NCTM Quantitative Literacy Project: An Overview and Possible Extensions." See also the report of the Focus Group on Quantitative Literacy in this volume.)

2. *Chance.* (J. Laurie Snell, Dartmouth)

This course is being developed at six colleges: Dartmouth, Grinnell, King, Middlebury, Princeton, and Spelman. The course studies chance issues currently in the news such as statistical issues in AIDS, gender issues in SAT examinations, the use of DNA fingerprinting in the courts, scoring streaks and records in sports, reliability of current political polls, interpreting data graphics, and the role of cholesterol in the prevention of heart disease. Such topics form the focus of the course. Concepts in probability and statistics are developed only to the extent necessary to understand the issues. Students read and compare treatments in the newspapers, popular science journals such as *Chance Magazine*, *Science*, and *Nature*, and original research articles. The goal is to make students more literate in probability and statistics and to permit them to make more intelligent choices when faced with chance issues. Experimental versions of the course are being given in the 1991 Fall term at Grinnell College by Tom Moore, at Middlebury College by William Peterson, and at Princeton University by Peter Doyle and Laurie Snell. These courses incorporate an emphasis on good writing, group learning, the use of the computer, and student projects.

3. *A Quantitative Reasoning Course.* (George Cobb, Mount Holyoke College)

In 1982, at Mount Holyoke College, a group of faculty began to plan what was eventually to find its way into the course catalog as Interdepartmental 100—Case Studies in Quantitative Reasoning. After gaining provisional approval by the faculty in 1986, the course was taught for the first time in 1987. As of fall 1990, the QR course, as it has come to be called, had been taught for eight consecutive semesters all told. Fifteen faculty from six departments had by then taught 22 sections of the course to a total of more than 250 students. The course, which has no prerequisites, is designed to appeal to students with a broad range of academic interests and widely differing mathematical backgrounds. It is not a simple presentation of technical methods followed by practice problems. Instead, case studies from a variety of disciplines form the subject matter of the course. Different quantitative methods are introduced and used in the attempt to develop understanding of these examples. The emphasis is not on rote computation, but on reasoning; not on formulas, but on ways to construct and evaluate arguments. The goals are to help students strengthen their analytical skills and acquire more confident understanding of the meaning of numbers, graphs, and other quantitative summaries they will encounter in many subsequent courses, no matter what their majors.

For example, witchcraft in seventeenth-century New England forms the central problem for investigation in the first third of the course, which concentrates on exploratory data analysis, with heavy emphasis on graphs, two-way tables, and other informal statistical tools for finding and presenting patterns in numerical data. In this case the data pertain to the 141 people accused of witchcraft in 1692 in Essex county, and to residents of Salem Village whose names appear in various tax records, petitions to the General Court, and minutes of the Village meetings. Discussions and assignments stress the process of working from numerical patterns to plausible explanations, and vice-versa, from possible explanations of the historical phenomena to relevant numerical evidence. The major project is to write a paper formulating, investigating, and discussing some hypothesis about the relationship between wealth and power as reflected in the historical records, along with some other hypothesis of the student's own choosing. ("The Quantitative Reasoning Course at Mount Holyoke College," in Samuel Goldberg (Ed.), "The New Liberal Arts Program: A 1990 Report," Alfred P. Sloan Foundation.)

4. *Statistical Thinking and Statistical Methods.* (Gudmund Iversen, Swarthmore College)

At Swarthmore we have carried this idea [Jim Landwehr's distinction between doing and thinking] so far that we have two introductory courses. In Statistical Methods I expect the students to be able to do their own analyses after they have finished the course. Most students end up doing the statistics in papers, lab reports, and senior theses. But this means going through regression, analysis of variance, and contingency tables. This year I am using the Moore/McCabe book (*Introduction to the Practice of Statistics,* Freeman, 1989). In Statistical Thinking I do not expect students to be able to do anything, but the goal is to have them understand uses of statistics they see all around them in scientific journals, books, newspapers, television, news magazines, etc. I am using the Moore paperback (*Statistics: Concepts and Controversies, 3rd edition,* Freeman, 1991) with additional material for this course. (A more detailed description appears as Appendix B.3.)

5. *Time Series Analysis As a First Course in Statistics.* (Robin Lock, St. Lawrence University)

A wide variety of fundamental statistical ideas can be conveyed through the study of time series. For example, the general notions of an underlying model for some real world phenomenon, estimation of its parameters from data, and diagnostic checking of the model assumptions are central themes in statistics. The models encountered in forecasting are fairly straightforward, yet can be used to effectively illustrate important principles such as parsimony, variability in parameter estimates, and criteria for choosing between competing models. The analysis of residuals to check model assumptions, suggest alternate models, or gauge the accuracy of the fit is a featured part of time series methodology which is often neglected in traditional introductions to statistics. Similarly, statistical graphics are used at many points throughout a time series analysis.

The point here is not to demonstrate that the field of time series analysis uses important statistical techniques, but rather that many of the fundamental concepts in applied statistics can be effectively introduced to students within the context of time series analysis. (Adapted from "Forecasting/Time Series Analysis: An Introduction to Applied Statistics for Mathematics Students," SLAW Technical Report No. 90-001. This and other technical reports of Statistics and the Liberal Arts Workshop are available from Don Bentley, Mathematics, Pomona College.)

6. *Multivariate Descriptive Statistics.* (Frank Wolf, Carleton)

A course in multivariate descriptive statistics that presupposes preparation in linear algebra can be used as a course to introduce students to statistical ideas. Such a course requires students to make heavy use of mathematical modelling and of concepts, techniques, and results from linear algebra and, hence, contributes to the students' understanding and appreciation of applied mathematics. The course is data-driven, and students learn to deal sensibly with real data. The course should be acceptable to a typical liberal arts mathematics department for credit in mathematics. It can be taught so as to assume no earlier training in statistics and yet be open and very useful to those students who have already taken one or more courses in that area. Such a course has been taught many times at Carleton. (From "Multivariate Descriptive Statistics: An Alternative Introduction to Statistics," SLAW Technical Report No. 90-004.)

7. *Data Analysis in the Mathematical Statistics Course.* (Jeff Witmer, Oberlin)

I believe it is imperative that students learn something of how statistical theory is applied in practice, but it is particularly difficult to cover much material on applied statistics while at the same time covering the mathematical statistics topics. I address this problem by offering an additional, one-credit course at the upper level.

In the spring of 1988, I started the course by discussing some of the statistical packages that are available on computers. I used MINITAB. I then presented some ideas from exploratory data analysis ... the use of normal probability plots and transformations of data ... compound smoothers ... statistical quality control ... I devoted roughly half the course to the general topic of regression, relying heavily on MINITAB.

In the spring of 1989, rather than present lectures, I involved the entire class in a data analysis project. We analyzed the results of a survey I helped conduct of libraries at liberal arts colleges. The students helped me explore roughly 200 variables measured on 97 colleges. I gave each of the students access to the computer file that contained the data and told them, "Explore, generate graphs, fit models, and let me know what you learn." We spent class periods discussing the data set and the statistical methods we were using to analyze it.

I am planning to teach the course in 1990 in a similar fashion. The reactions of students to this course have been positive. I believe, and they seem to agree, that seeing statistical methods applied to real data motivates students to want to learn more about the subject. (From "Data Analysis: An Adjunct to Mathematical Statistics at Oberlin College," SLAW Technical Report No. 90-003.)

8. *Computer-Enhanced Mathematical Statistics.* (Marsha Davis, Mount Holyoke)

Under a grant from FIPSE (Fund for Improvement of Post-Secondary Education), Mathematical Statistics at Mount Holyoke College has been redesigned, and now meets in a computer classroom. With the aid of technology, the new course incorporates a constructivist approach to mathematics instruction. Laboratory projects have been designed to support theoretical material and to guide students in discovering concepts for themselves. In one project, for example, students use computer simulation to examine the sampling distributions of sample means as an introduction to the normal distribution and the central limit theorem. Another project, "The Taxi Problem," places students in a hypothetical situation where they must estimate the parameter N of a discrete uniform distribution on the integers $1 - N$. Students suggest plausible estimators, generate ideas of reasonable criteria for selecting a "best" estimator, modify estimators based on theoretical considerations, and present their choice of estimator based on results from a simulation study. Through this process students gain an understanding of how a research statistician works, as well as a chance to experience the interplay of working with theory and testing ideas with simulation.

Foster Active Learning

The following examples correspond to the various alternatives to lecturing mentioned in our third recommendation: group problem solving and discussion, lab exercises, demonstrations, written and oral presentations, and projects. As many of the examples illustrate, there are effective alternatives to lecturing that do not use computers. (See also Robin Lock and Tom Moore, "Low-Tech Ideas for Teaching Statistics," SLAW Technical Report No. 91-008.)

On the other hand, certain kinds of lab exercises (e.g., Nos. 3, 7, 8 above) and class demonstrations (No. 3b below) are impossible without computers, as are most projects (Nos. 5a and 5b below). In particular, demonstrations that rely on even the simplest dynamic graphics (No. 3b below) can be extremely effective, and are impossible without computers.

1. *Group Problem Solving and Discussion.* (Joan Garfield, University of Minnesota)

In my classes, I do not lecture at all, which takes a while for students to adjust to. Instead, students are required to read the textbook before coming to class, guided by a study guide/student handbook I have written containing study questions, sample problems, etc. When students come to class each day we first discuss the study questions, often arguing about issues such as which is the best measure of center to use, which type of plot gives the most information, etc. They rapidly learn that there is often not one right answer nor one way to solve a problem. At first they think this means that anything is OK, but then learn what is important is being able to justify a claim, defend a point of view, and judge the appropriateness of a solution process. After our large group discussions, students work in permanent small groups on activities, usually analyzing a set of data and discussing questions about these data sets. I find the Quantitative Literacy Project materials very useful for these small group activities. Each write-up is turned in that day and the group receives a group score. I also give weekly quizzes on the homework problems. This method works extremely well. Students tend to enjoy the course and express amazement that they are actually doing statistics (and it's fun!). Many bring in data sets of interest to them to analyze for their "real life" problem assignments, and learn how to deal with messy data that are not easily plotted, missing values, and related issues. (A more detailed description appears as Appendix B.2.)

2. *Lab Exercises.* (Dick Gunst, Southern Methodist University)

I have a "nuts and bolts" experiment that I've used many times to illustrate (a) the need for data, (b) application of a simple two-factor factorial experiment, and (c) the ability of simple graphics (point plots) to convey important information. The experiment involves students selecting nuts and bolts from a tall, small mouth jar containing a variety of sizes of nuts and bolts. The objective is to select and screw four nuts onto four bolts in as short a time as possible. Students operate in groups, taking turns selecting and fastening the nuts and bolts and timing those who do. After the initial times are obtained and plotted, discussion ensues over why the times are unacceptable and what could cause a reduction of the times. A simple experiment is then conducted using four combinations of jar sizes and nut sizes and the results are again plotted. The sorting of jars and nuts results in lower times *for all four groups*, an unexpected result. (A more detailed description appears as Appendix B.4.)

3a. *Demonstrations Based on Self-Generated Time Series.* (Howard Taylor, Delaware)

(There are no prerequisites for this one.) The students are asked to write down a random series of $n = 100$ numbers chosen from the digits 1, 2, ... , 20. The form provided has columns with 1. ___, 2. ___, etc., to 100. ___, a psychological nudge to write the numbers down sequentially. (But some of the brighter students will skip around, entering a bunch of 1's, then a bunch of 2's, and so on.) For comparison, one or more students (or the instructor) uses a table of random numbers, or a programmable hand calculator, to form their lists.

The analysis:

(i) The students are asked to do a frequency tabulation and draw a histogram. The result is that most students have frequencies that look fairly realistic.

(ii) Next, a time series plot is drawn. These typically look less realistic when compared to the control plot. Some students go up-down too much; others are too smooth.

(iii) This suggests a plot of $X(t)$ vs $X(t+1)$ which typically looks less random, showing either a positive or negative correlation.

(iv) Students are asked to count how many times $X(t) = X(t+1)$, and how often this should happen in a random time series.

(v) Students are asked to calculate the number of runs above and below the median R. The critical points for a test of randomness are $R \leq 40$ or $R \leq 62$. These critical points are written on a slip of paper and given to a student prior to the calculations of the R's. Then the time series are divided into two groups according to the R values. This virtually always very nicely divides them into human generated versus computer or table generated, effectively demonstrating how hard it is for humans to be truly random.

(Two more examples of demonstrations appear in Appendix B.1.)

3b. *Simple Computer-Based Demonstrations.* (Ann Watkins, California State University at Northridge)

(i) Standard deviations: One computer demonstration I use in class requires a large screen projector but only the simplest software. Students can construct a population of, say, 100 numbers between 1 and 100. The program computes the mean and standard deviation. The game we play in class is to find the population with the largest standard deviation (and then with a given standard deviation). This sounds incredibly simple-minded, but students find it challenging as they are just coming to grips with the idea of standard deviation.

(ii) Regression: You can build effective labs or class demonstrations using programs that allow students to change the points of a scatterplot and watch the least squares line (and coefficients) change, or leave the points fixed but change the fitted line, and watch the residual sum of squares change.

4. *Written and Oral Presentations.* (Gudmund Iversen, Swarthmore College)

There is an increased emphasis on writing in today's undergraduate curriculum, and papers can play an important role in an introductory statistics course. With the existence of good interactive statistical software it is possible to move the classic introductory statistics course away from the study of formulas to the study of statistical thinking and the role of statistics in society. In such a new course students get an increased understanding of statistical ideas by writing papers across a wide range of topics; actual topics have ranged from a comparison of statistics and religion to a study of the relationship between the time of first class in the morning and the distance from the bed to the alarm clock. (See "Writing Papers in a Statistics Course," Proceedings of the Section on Statistical Education, American Statistical Association, 1991. See also Norean Radke-Sharpe, "Writing As a Component of Statistics Education," *American Statistician,* Vol. 45, No. 4, November 1991.)

5a. *Projects.* (Tom Moore, Grinnell, and Katherine Halvorsen, Smith)

Student projects can teach concepts not usually encountered in introductory or second-level statistics courses. Questions about study design, study protocols, questionnaire construction, informed consent, confidentiality, data management, data cleaning, and handling missing data may arise when students deal with collecting and analyzing their own data. ... This paper describes the student projects we have used in introductory courses and in second-level statistics classes. It addresses the issues of motivating, monitoring, and evaluating student projects, and discusses some unique problems student projects present for instructors using them. ... In our experience students usually conclude that the project was one of the most useful parts of the course. Some comment that the project made them apply everything they learned as soon as they learned it. For some it is the first time they have stood in front of an audience to present their work. On the whole we would encourage other faculty to use this kind of project in their classes. (See "Motivating, Monitoring, and Evaluating Student Projects," Proceedings of the Section on Statistical Education of the American Statistical Association, 1991.)

5b. *Projects.* (Don Bentley, Pomona)

There are at least three types of exercises involving data for students to work with at the introductory level: the standard fictional data set, real data extracted from the literature, and statistical consulting. Students who have a potential interest in statistics as a career should be given the opportunity to become meaningfully involved in the analysis of data from original scientific investigations ... to encounter the excitement of being the first to know the results of the statistical analyses. ...

A list of projects with which students have been involved include soaking and cleaning solutions for hard contact lenses, sterilizing solutions for soft contact lenses, proving efficacy of intraocular lenses, evaluation of a retroprofusion process used with angioplastic surgery, and investigation of the use of ambulatory tocodynamometry in high risk pregnancies. These industry-generated projects provide a wealth of opportunities for students to become involved in the analysis of meaningful data. The necessity for accuracy is clearly defined. This is an element frequently lacking in the classroom experience. The importance of the role of statistics in the research is also made very clear to the student, both by the financial implications of the project to the corporation, and the implications for health care in general. (The consequences of both the Type I and Type II errors gain real meaning.) (See "Recruiting and Training Undergraduates Through Statistical Consulting," Proceedings of the Section on Statistical Education, American Statistical Association, 1991.)

Making It Happen

The way faculty change and learn is probably not so very different from the way students change and learn. In particular, reports alone probably have about as much effect on the way most faculty teach statistics as lectures alone have on the way most students understand statistics. Our Focus Group may be full of sound and fury, but without effective follow-through, we'll fail to signify even at the ten percent level.

Change must overcome four inertias: one logistical, one intellectual, one interpersonal, and one institutional.

1. Logistical Inertia:

Good data sets are hard to find. Anyone reading this report could easily invent enough examples to fill a lecture on differentiating polynomials, and it would take at most five minutes or so. But how long would it take you to come up with just one real data set, say from cognitive psychology, to illustrate the effect of outliers in one-way analysis of variance?

Automating statistical busy-work has a high start-up cost. Unless you're already set up with a data analysis package for your classes, the ordeal of first choosing a good one, then getting it installed, then learning how to use it yourself, and finally learning how to teach your students to use it—all this may not be at the top of your list of ways to spend what would otherwise be your next vacation.

2. Intellectual Inertia:

Learning to handle the ambiguities of statistics takes time, practice, and hard thought. Even with software installed and data sets in hand, doing a proper analysis and interpretation is a kind of challenge that many who teach statistics are not prepared to meet, mainly because, through no fault of their own, they've rarely if ever seen it done, and their training and experience have not prepared them either to do it or to value the doing of it. Remember that in Plato's curriculum, students were to devote their entire first decade of study to mathematics, because the other subjects, being less clear-cut, were understood to be *harder.* (Mathematics is often mistaken for being harder because the absence of ambiguity makes the subject much less forgiving of low quality effort.)

3. Interpersonal Inertia:

Giving up the familiar role of "I talk, you listen" doesn't always come easily. Teachers who are skillful and comfortable delivering a lecture on volumes of revolution may, through lack of role models and lack of practice, find themselves neither skillful nor comfortable with the group dynamics of labs or discussions, where the direction of conversation is not set in advance, and there will often be more than one defensible position on what the data have to say.

4. Institutional Inertia:

Most deans and department heads don't care very much whether statistics is taught well. One consequence is that untenured mathematics faculty are all too often being quite realistic to see a fork in their career path—to the left, doing what it takes to teach statistics well, or to the right, doing what it takes to get tenure. A second consequence, apparently, is that many mathematics departments have not been willing to do what it takes to recruit and retain a Ph.D. statistician. According to Moore and Roberts (*American Statistician*, 1989, 43:2) only 26 of 80 mathematics departments at responding liberal arts colleges could claim a Ph.D. statistician. Half had no one with even a masters-level degree in statistics.

In the hope of overcoming these inertias, we offer two clusters of meta-recommendations:

Meta-Recommendation I: WORK AT THE GRASS ROOTS

The statistics and mathematics professions should do more to support those who want to teach statistics well, through (1) new sections in journals, (2) a newsletter, (3) improved access to data, (4) more e-mail conferencing, (5) more workshops for teachers, and (6) better preparation for teaching in graduate schools.

It seems to me that ASA could do considerably more about undergraduate teaching than it currently does. For example, the other ASA to which I belong (American Sociological Association) maintains a Teaching Resource Center that collects and circulates such teaching materials as syllabi and reading lists and holds teaching workshops. The Association also sponsors a journal entitled *Teaching Sociology* that is a veritable treasure chest of ideas for managing classes and of research on the efficacy of teaching strategies.

—*Judy Tanur, SUNY Stony Brook*

1. *Establish New Sections in Existing Journals.*

It would be easy to assume that most of the things we recommend should be done by the ASA, but some of this must be done in MAA journals. Quite a few statistics teachers in mathematics departments, members of MAA but not ASA, need to be encouraged to participate in this conversation.

—*Mary Parker*

There should be sections devoted to statistical education in some of the existing journals. A section on statistical education devoted to undergraduate teaching and learning should not be used as a vehicle to publish low quality research papers. I'm opposed to starting a new journal. Libraries are cutting circulation and I can't find time to read existing journals.

—*Jack Schuenemeyer*

The "Teacher's Corner" in the *American Statistician*, although interesting, does not address our concerns.

—*Mary Parker*

I'd like things set up so most submissions would be short, each one describing a project or lab or class demo. There should be a well-planned classification scheme, so that a highly structured index would be easy to construct, to make it possible for someone wanting a demo on the central limit theorem for a large lecture class to locate quickly a write-up of Howard Taylor's example; ditto for someone wanting examples of two-hour computer labs on regression diagnostics for a small class of students with no linear algebra background, or . . .

—*George Cobb*

2. *Establish a Newsletter.*

I like the idea of a newsletter to be sent out three or four times a year. The "Undergraduate Mathematics Education" newsletter (*UME Trends*) began as a funded project (NSF, I think) and now is supported by paid subscribers like myself. It is an excellent source of information about current projects, programs, research, books, software, conferences, workshops, teaching ideas, etc., all related to teaching undergraduate mathematics courses.

—*Joan Garfield*

3. *Expand Efforts to Make Data Sets Readily Available.*

Statisticians should contribute more often to the bank of data sets available by e-mail via the StatLib file server at Carnegie Mellon University. (See Appendix C for a brief description and instructions on how to get data sets.) Data sets should be classified and indexed by area of application, structure (e.g., experimental two-way factorial in complete blocks), and interesting statistical features (e.g., data need to be transformed to logs), so that teachers could use the index to retrieve examples to fit particular pedagogical needs. A newsletter or section of a journal could list such summary information for data sets that had been contributed since the last issue. Periodically, anthologies of these data sets should be published.

4. *Expand Efforts to Involve Statistics Teachers in E-mail Conferencing.*

Tim Arnold of North Carolina State has just recently set up a procedure for joining such a group, one established by the Statistical Education section of the ASA. (See Appendix C for a description and instructions on how to join.)

> In the ten years or so I have been teaching statistics, I have not seen many opportunities to participate in the kind of discussion we are having in this group. I think that statistics teachers would profit by, and enjoy, participating in such discussions. ...We need to be providing the kind of forum and inspiration to people that will encourage each of them to think of ideas of their own. For me, reading what other people do is marvelously inspiring. I think it would be to others.
>
> —*Mary Parker*

5. *Run Workshops at National and Regional Meetings.*

6. *Establish Programs Within University Departments to Help Prepare Graduate Students to Teach.*

> Our department also has a formal program for teaching our graduate students to teach. (Of course, what they teach is introductory sociology, not statistics, but I would think some of the ideas are transferable.) In the fall semester the teaching practicum meets once a week to study the art and craft of teaching, observe teachers who have a reputation for expertise and discuss their techniques with them, etc. Each member of the practicum prepares a syllabus for an introductory course, and gives several practice classes to the other members of the practicum. In the spring semester, each graduate student teaches his/her own section of the introductory sociology course under the supervision of the practicum instructor (often me). Again, it seems to work. The graduate students are considerably less nervous and considerably more effective with their classes than they would otherwise have been, and they have won far more than their share of university-wide awards for excellence in teaching. But again, this system is very demanding of resources.
>
> —*Judy Tanur*

Meta-Recommendation II: WORK AT THE TOP

Statisticians and mathematicians should work together with academic administrators—first to improve the way colleges evaluate work by statisticians and others who teach statistics, and second, to insure that improved evaluation reshapes institutional criteria both for faculty recruitment and for subsequent personnel decisions.

> Without additional motivation and encouragement, I don't think our material will be read by those who need it most. That kind of motivation usually has to come from one's senior colleagues within a department. What if young faculty see that the reward system is weighted mostly toward refereed publications? We need to enlist the help of others who will influence them. The obvious group is Department Heads. That could be done through the auspices of the MAA and ASA, both of whom do have periodic Department Heads meetings.
>
> —*Walt Pirie*

> We need to think hard about how to reach not just teachers of statistics, but also their chairmen who assign them the statistics courses. We need to impress presidents and provosts and deans that

things are bad, that they need to do something about the teaching of statistics in their schools, and that we are available to help.

—Gudmund Iversen

The comment which follows, by a newly elected Fellow of the ASA, came early in our group's work, but serves well as a final summary:

> At Delaware we teach an introductory, non-calculus based, traditional, two-semester course to 1000 students per year. Problems: too much lecturing, too much emphasis on hypothesis testing, too much material, too little involvement by students, and too little reward for those who spent time trying to improve the situation.

—Jack Schuenemeyer

Focus Group Participants

bushaw@wsuvm1.bitnet	DONALD W. BUSHAW, *Washington State University.*
gcobb@mhc.bitnet	GEORGE W. COBB, Moderator, *Mount Holyoke College.*
gunst@smuvm1.bitnet	RICHARD F. GUNST, *Southern Methodist University.*
iversen@swarthmr.edu	GUDMUND IVERSEN, *Swarthmore College.*
jml@research.att.com	JAMES M. LANDWEHR, *AT&T Bell Laboratories.*
dsm@chi.stat.purdue.edu	DAVID S. MOORE, *Purdue University.*
mooret@grin1.edu	THOMAS L. MOORE, *Grinnell College.*
parker@math.utexas.edu	MARY R. PARKER, *Austin Community College.*
wltpirie@vtvm1.edu	WALTER R. PIRIE, *Virginia Polytechnic Inst. & State Univ.*
scheaffe@orca.stat.ufl.edu	RICHARD L. SCHEAFFER, *University of Florida.*
jacks@brahms.udel.edu	JOHN H. SCHUENEMEYER, *University of Delaware.*
jtanur@sbccvm.sunysb.edu	JUDITH M. TANUR, *SUNY Stony Brook.*
pamh@brahms.udel.edu	HOWARD M. TAYLOR, *University of Delaware.*
awatkins@vax.csun.edu	ANN E. WATKINS, *California State University at Northridge.*
pqa6031@ca.acs.umn.edu	JOAN GARFIELD, *University of Minnesota.*

Suggested Reading

EXPLORATORY DATA ANALYSIS

1. Hoaglin, David C.; Mosteller, Frederick; Tukey, John W. (Eds.). *Exploring Data Tables, Trends, and Shapes.* New York: John Wiley, 1985.
2. Hoaglin, David C.; Mosteller, Frederick; Tukey, John W. (Eds.). *Understanding Robust and Exploratory Data Analysis.* New York: John Wiley, 1983.
3. Mosteller, Frederick and Tukey, John W. *Data Analysis and Regression.* Reading, MA: Addison-Wesley, 1977.
4. Tukey, John W. *Exploratory Data Analysis.* Reading, MA: Addison-Wesley, 1977.

GRAPHICS

1. Becker, Richard A.; Cleveland, William S.; Wilks, Alan R. "Dynamic graphics for data analysis," *Statistical Science*, 1986, V. 1, pp. 355–395.

2. Chambers, J.M.; Cleveland, W.S.; Kleiner, B.; Tukey, P.A. *Graphical Methods for Data Analysis*. Belmont, CA: Wadsworth, 1983.

3. Cleveland, William S. *The Elements of Graphing Data*. Belmont, CA: Wadsworth, 1985.

4. Cleveland, William S. and McGill, M.E. (Eds.). *Dynamic Graphics for Statistics*. Belmont, CA: Wadsworth, 1988.

5. Kolata, Gina. "Computer graphics comes to statistics." *Science*, 217 (1982), p. 919.

6. Tufte, Edward R. *The Visual Display of Quantitative Information*. Chester, CT: Graphics Press, 1983.

7. Tukey, John W. "Data-based graphics: Visual display in the decades to come." *Statistical Science*, 5:3 (1990), pp. 327–339.

8. Wainer, Howard and Thissen, David. "Plotting in the modern world: Statistical packages and good graphics," *Chance*, 1:1, 1988.

9. Weihs, Claus and Schmidli, Heinz. "On-line multivariate exploratory graphical analysis." *Statistical Science*, 5:2 (1990), pp. 175–226.

MISCELLANEOUS

1. Barnett, Vic. *Comparative Statistical Inference, Second Edition*. New York: John Wiley, 1982.

2. Belsley, D. A., Kuh, E., and Welsch, R. E. *Regression Diagnostics: Identifying Influential Data and Sources of Collinearity*. New York: John Wiley, 1980.

3. Cook, R. Dennis and Weisberg, Sanford. *Residuals and Influence in Regression*. London: Chapman and Hall, 1982.

4. del Pino, Guido. "The unifying role of iterative generalized least squares in statistical algorithms." *Statistical Science*, 4:4 (1990), pp. 394–408.

5. Efron, Bradley and Tibshirani, R. "Bootstrap methods for standard errors, confidence intervals, and other measures of statistical accuracy." *Statistical Science*, 1 (1986), pp. 54–77.

6. Hoaglin, David C. and Moore, David S. (Eds.). *Perspectives in Contemporary Statistics*. Washington, DC: Mathematical Association of America, 1992.

7. Walton, Mary. *The Deming Management Method*. New York: G.P. Putnam, 1987.

SOME WAYS STATISTICS DIFFERS FROM MATHEMATICS

1. Kempthorne, Oscar. "The teaching of statistics: Content versus form." *The American Statistician*, 34 (1980), pp. 17–21.

2. Landwehr, J.M. "Discussion of the role of statistics at four-year undergraduate institutions." *Proceedings of the Section on Statistical Education of the American Statistical Association*, 1990.

3. Moore, David S. "Should mathematicians teach statistics?" (with discussion). *The College Mathematics Journal*, 19 (1988), pp. 3–25.

4. Moore, David S. "Teaching statistics as a respectable subject." SLAW Technical Report No. 91-002, Department of Mathematics, Pomona College.

5. Moore, Thomas L. and Witmer, Jeffrey. "Statistics within departments of mathematics at liberal arts colleges." *American Mathematical Monthly*, 98 (1991), pp. 431–436.

Appendix A: Helping Students Learn

by Joan Garfield, UNIVERSITY OF MINNESOTA

A summary of research related to learning statistics, drawn from the areas of learning and cognition, mathematics education, and statistical learning:

A. Basic Concepts are Hard, Misconceptions Persistent

1. *Teachers should not underestimate the difficulty students have in understanding basic concepts of probability and statistics.* Ideas of probability and statistics are very difficult for students to learn and conflict with many of their own beliefs and intuitions about data and chance (Shaughnessy, in press; Garfield and Ahlgren, 1988). Students may be able to answer some test items correctly or perform calculations correctly while still misunderstanding the basic ideas and concepts (Konold, 1990; Garfield and delMas, in press). Students' misconceptions may be strong and resilient (Konold, in press; Garfield and delMas, in press; Shaughnessy, in press). "It is very difficult to replace a misconception with a normative conception, a primary intuition with a secondary intuition, a judgmental heuristic with a mathematical model. Beliefs and conceptions are slow to change" (Shaughnessy, in press).

2. *Learning is enhanced by having students actually become aware of and confront their own misconceptions.* Activities that help students evaluate the difference between their own beliefs and intuitions about chance events and actual empirical results help students learn even better (delMas and Bart, 1989; Shaughnessy, 1977). If students are first asked to make guesses or predictions about data and random events, they are more likely to be engaged and motivated to learn the results. When experimental evidence explicitly confronts their predictions, they should be helped to evaluate this difference. In fact, unless students are forced to record and then compare their predictions with actual results, they tend to see in their data confirming evidence for misconceptions of probability. Research in physics instruction also points to this method of testing beliefs against empirical evidence (e.g., Clement, 1987).

B. Learning is Constructive

1. *Learning is a constructive activity.* Students learn by constructing knowledge, not by passive absorption of information (Resnick, 1987; von Glasersfeld, 1987). Students approach a learning activity with prior knowledge, assimilate new information, and construct their own meaning (NCTM, 1989). They connect the new information to what they already believe. Students accept new ideas only when their old ideas don't work or are inefficient. Ideas are not isolated in memory but are organized and associated with language and experiences. Students have to construct their own meaning for what they are learning, regardless of how clearly a teacher or a book tells them something (AAAS, 1989). Students retain best the mathematics that they learn by processes of internal construction and experience (NRC, 1989).

2. *Students learn better by active involvement in learning activities.* Students learn better if they are engaged and motivated to struggle with their own learning. Teachers should involve students in their own learning by using activities that require students to express their ideas both orally and in writing. Students learn better if they work cooperatively in

small groups to solve problems and learn to argue convincingly for their approach among conflicting ideas and methods (NRC, 1989).

3. *Students learn to do well only what they practice doing.* Practice may be through hands-on activities, cooperative small groups, or computer work. Students also learn better if they have experience applying ideas in new situations. If they practice only calculating answers to well-defined problems, then that is all they are likely to learn. Students cannot learn to think critically, to analyze information, to communicate ideas, to make arguments unless they are permitted and encouraged to do those things over and over in many contexts. Merely repeating and reviewing tasks is unlikely to lead to improved skills or deeper understanding (AAAS, 1989).

4. *Effective learning requires feedback.* Learning is enhanced if students have opportunities to express ideas and get feedback. Feedback ought to be analytical, and come at a time when students are interested in it. There must be time for students to reflect on the feedback they receive, make adjustments, and try again (AAAS, 1989).

5. *Computers should be used to enhance learning, not just to crunch numbers.* Computer-based instruction appears to help students learn basic statistics concepts by providing visual representations as well as the capability for easy interactive data exploration (Rubin, Rosebery and Bruce, 1988; Weissglass and Cummings, in press).

6. *Statistics instruction should shift from lecture-based classes emphasizing learning mechanical skills, to student-centered courses emphasizing application, understanding, and communication.* Based on research in learning and cognition, and mathematics education, the National Council of Teachers of Mathematics (1991) is urging the following shifts in mathematical instruction, which also apply to statistics instruction:

- Toward classrooms as mathematical communities.
- Toward logic and mathematical evidence as verification, away from the teacher as the sole authority for right answers.
- Toward mathematical reasoning, away from merely memorizing procedures.
- Toward conjecturing, inventing, and problem solving, away from an emphasis on mechanistic answer-finding.
- Toward connecting mathematics, its ideas, and its application away from teaching mathematics as a body of isolated concepts and procedures.

References

1. American Association for the Advancement of Science. *Science for all Americans.* Washington, DC: American Association for the Advancement of Science, 1989.
2. Clement, J. "Overcoming Students' Misconceptions in Physics: The Role of Anchoring Intuitions and Analogical Validity." Proceedings of the Second International Seminar, Misconceptions and Educational Strategies in Science and Mathematics. Ithaca, NY: Cornell University, 1987.
3. Garfield, J.B. and Ahlgren, A. "Difficulties in Learning Basic Concepts in Probability and Statistics." *Journal for Research in Mathematics Education,* 19 (1988), 44–63.
4. Garfield, J.B. and delMas, R. "Students' Conceptions of Probability." In *Proceedings of the Third International Conference on Teaching Statistics.* Dunedin, New Zealand. (In press.)

5. Konold, C. "The Origin of Inconsistencies in Probabilistic Reasoning of Novices." In *Proceedings of the Third International Conference on Teaching Statistics,* Dunedin, New Zealand. (In press.)

6. National Council of Teachers of Mathematics. *Curriculum and Evaluation Standards for School Mathematics.* Reston, VA: National Council of Teachers of Mathematics, 1989.

7. National Council of Teachers of Mathematics. *Professional Standards for Teaching Mathematics.* Reston, VA: National Council of Teachers of Mathematics, 1991.

8. National Research Council. *Everybody Counts: A Report to the Nation on the Future of Mathematics Education.* Washington, DC: National Academy Press, 1989.

9. Resnick, L. "Education and Learning To Think." Washington, DC: National Research Council, 1987.

10. Rubin, A.; Rosebery, A.; and Bruce, B. "ELASTIC and Reasoning Under Uncertainty." Research Report No. 6851. Boston, MA: BBN Systems and Technologies Corporation, 1988.

11. Shaughnessy, J.M. "Misconceptions of Probability: An Experiment With a Small-group Activity-based Model Building Approach to Introductory Probability at the College Level." *Educational Studies in Mathematics,* 8 (1977), 285–316.

12. Shaughnessy, J.M. "Research in Probability and Statistics." In D.A. Grouws (Ed.), *Handbook of Research on Mathematics Teaching and Learning.* New York: Macmillan, 1977.

13. Von Glasersfeld, E. "Learning As a Constructive Activity." In C. Janvier (Ed.), *Problems of Representation in the Teaching and Learning of Mathematics.* Hillsdale, NJ: Lawrence Erlbaum Associates, 1987, pp. 3–17.

14. Weissglass, J. and Cummings, D. "Dynamic Visual Experiments with Random Phenomena." In Walter Zimmerman and Steve Cunningham (Eds), *Visualization in Teaching and Learning Mathematics.* MAA Notes No. 19. Washington, DC: Mathematical Association of America, 1991.

Appendix B: Examples

1. Classroom Experiments (Howard Taylor)

Central Limit Theorem. (Prerequisite: Probability distribution, means and variances; a class of 40 to 100.) I start with a jar containing tags numbered 0, 1, ..., 9 and select a tag at random, writing its number on the board. Then I select ten tags, with replacement, and compute their sample mean. A student repeats the experiment, providing another single digit and another sample mean. Then in the interest of speed, I pass out slips of paper, each slip containing a single digit plus ten more digits and their sample mean. (I create these using APL.) The students call out their single digit in order and we create a frequency tabulation. Usually you have to smooth it into five groups to get a reasonable histogram. You can compare the observed frequencies with the uniform probabilities. Of course the sample experience does not exactly conform to the population of the sample means. Prior to the tabulation, I predict the frequencies in the various groups from normal table calculations that I make ahead of time. From the histogram, we observe that the sample means are centered about the same number as the individual observations, there is less variability in the sample means, and the bell-shaped curve has started to appear. Again, we compare the observed frequencies to the predicted frequencies. (The students are often amazed that I can make these predictions.)

Acceptance Sampling. (Prerequisite: Elementary binomial probabilities.) We compute the probability of accepting the lot when a sample size of 20 (and 50) items is selected, and we accept the lot if there are c or less defectives in the sample, as a function of the fraction p of defectives in the very large lot. We draw the operating characteristic curves for $c = 0, 1, 2$ and $0 < p < .20$. We then choose $n = 50$ and $c = 1$ and pass out simulated data, 5×10 arrays of 0's and 1's representing each student's sample. The first data is from a low acceptable value of p. We count how many student's accepted their lots, and how many rejected theirs, and compare the fraction accepted with the operating characteristic curve. Then we repeat with simulated samples from a higher value of p.

I find acceptance sampling a natural way to introduce students to hypothesis testing. Obviously, I treat Type I and Type II errors right from the start. One of the things pointed out at the end is that we could use a similar simulation to evaluate a very complicated acceptance rule whose operating characteristics we could not compute analytically.

2. Teaching Without Lecturing (Joan Garfield)

Case Study: An Alternative Approach to Teaching Introductory Statistics. This course was developed to introduce undergraduate students to the basic statistical ideas and methods. The only prerequisite for the course is a high school level course in algebra. The course is based on the following components:

1. *Involvement with real data.* Students are involved in the analysis of real data in order to solve problems of interest to them. This involves analyzing existing data sets gathered from a variety of resources as well as collecting data of their own.

2. *Emphasis on exploring data.* The focus of the course is on exploring data and this theme is continued throughout the course. The intent is to help students develop the attitude that statistical work is like detective work: looking for clues in the data, speculating about the real situation that a sample of data represent, and learning how different ways of representing data give different information.

3. *Use of software designed to build conceptual understanding.* Software is used not just as tools to crunch numbers and to display data more quickly and easily than by hand, but as a means to build concepts, confront misconceptions, and explore data by creating multiple representations.

4. *Emphasis on oral and written communication.* Students are given opportunities to communicate about statistics both orally (as they solve problems and analyze data in groups) and in written form (writing interpretations of data and results of student projects).

5. *Confronting misconceptions.* Students are confronted with their faulty reasoning, incorrect intuitions, and misconceptions through carefully designed experiences on the computer, group problem solving, and class discussion.

Course Details. The text used is *Statistics and Data Analysis* by Andrew Siegel. A study guide, written by the instructor, offers additional information, study questions to guide students' reading of the text, and sample problems with full solutions. Students are instructed to read the assigned chapters, jot down answers to the study questions, and work out the sample problems before coming to class. The class meets three times a week for ten weeks. Two sessions are two hours long and the third meets for fifty minutes (where students turn in homework and take a quiz).

Each two-hour session begins with a discussion of the assigned material and questions students have about the reading and homework assignment. This takes about 30 minutes. The next 45 minutes students work in assigned groups of 3 to 4 students per group, analyzing one or more small data sets provided to them which use the methods from their assigned readings. These data sets are accompanied by questions which help students focus on what to look for in their analysis, and require them to summarize in writing what they did and what they learned.

The next 45 minutes are spent in a computer lab using software to illustrate techniques, explore data, build conceptual understanding, or confront misconceptions. The software programs currently used are not yet commercially available, but are available from the authors. They are the Chance Plus Software (Clifford Konold, SRRI, Hasbrouck Lab, University of Massachusetts, Amherst, MA 01003), and ELASTIC Software (Andee Rubin, TERC, 2067 Massachusetts Ave., Cambridge, MA 02140).

Evaluation. Students receive grades in the course based on their scores on each of the following: weekly homework assignments; weekly quizzes; daily group activities; mid-term and final exams; and two student projects (where students individually collect and analyze a set of data and write up a full report).

I collect informal evaluation information about the course in several ways:

1. I usually have students complete a mid-quarter survey which asks about their reactions to the course and its nontraditional format (i.e., no lectures, reading the book before coming to class, and working in groups on problems during class time). Their responses are 90% enthusiastic, with comments such as: "I never knew statistics could be fun," "I can't believe I actually enjoy this course," etc.

2. I use a standard student evaluation form at the end of the quarter which indicates generally favorable student reactions. I know this is not just a teacher effect because I have trained six different graduate students to use my materials and methods, and students have expressed high levels of satisfaction in those sections of the course as well, even with two foreign students who often get lower ratings in traditional lecture classes.

3. I use what I call "Real Life Problems." These are structured written reports of student projects where students have gathered and analyzed a set of data. Although this is used as a learning activity, it also provides a useful assessment of students' ability to use the statistical language appropriately, choose appropriate methods of analysis, interpret summary measures and graphs of data, and draw reasonable conclusions based on their analysis. As an assessment method, it helps me see in which areas students are weak and may need additional work. Usually I find some consistent errors that then lead me to discuss these in class or structure additional activities for students to work on in groups.

3. Two Kinds of Introductory Courses (Gudmund Iversen)

Not all students have the same statistical needs. Some students will be actively collecting data in laboratory experiments or in small sample surveys and analyzing these data for course projects and possible theses. Other students will be analyzing secondary data for similar purposes. A large group of other students will be exposed to statistical results and conclusions in their readings of books and journals for a variety of different courses as well as in their reading of newspapers and news magazines. But these students will never be asked to collect or analyze data on their own.

The premise of this dichotomy leads to the construction of two kinds of (non-calculus level) introductory courses. A possible name for the first type of course is Statistical Methods, and this is essentially the type of introductory course commonly taught across the land. A possible name for the second type of course is Statistical Thinking, and this is a much more unusual course, even though varieties of this course are beginning to appear.

Having tried this distinction for the last five years, my own evaluation is that it has been very successful. When only Statistical Methods was offered, the students were much more mixed in their expectations, their interest in the material, and in their abilities to do the work. Now that the students have a choice, the Statistical Methods course draws students who are motivated to learn statistics because they know they need these methods for their work in other courses. Sometimes this motivation comes late, and they take the course as late as their senior year. But by that time they are often involved with their own data, and they really see the need for knowing statistics. The enrollment in this course is now larger than it used to be, and it is now typically taken by about 20% of any graduating class.

The goal of the Statistical Thinking course is for students to understand and critically evaluate statistical results they are exposed to both in their academic work and in their non-academic lives. This means they have to know about data collection, analysis, and inference, but they do not need to have the technical knowledge necessary to do any of these themselves. In practical terms this means the students do not have to see and study formulas.

For example, instead of ever seeing the formula for the correlation coefficient r, students are shown scatterplots with different degrees of correlation and are told that we can compute a number which ranges from zero to one, the value depending on how close the points are to a line through the points. Giving the r-value for each plot and seeing the degree of scatter that goes with each r teaches non-mathematically inclined students more about the strength of relationships than any study of the formula for r can do.

In the study of data collection the need is stressed for proper statistical random samples and randomized experiments. The students quickly catch on to the fact that self-selected samples and convenience samples cannot be used for generalizations. In the data analysis part of the course the focus is on simplifying the data so we can understand what they tell us without the loss of too much information. Graphical methods, with the use of overheads showing good and bad graphs, is a popular topic. Computational methods are harder to deal with, but it is still possible to explain about mean and standard deviation without writing down formulas.

The inference part centers around p-values, both for tests for single variables and tests for relationships between variables. The technical details of whether we need to compute z, t, χ-square, or F are interesting for statisticians but only of minimal interest and concern for a person with interests in the substantive matter at hand. These quantities are only tools used to find the p-value, which tells the important story about the data. Now that computer software routinely supply us with p-values, a student needs only to understand the proper interpretation of this concept and to know where to look on the output to find the actual value.

The last section of the course consists of a discussion of the role of statistics in today's society. It stresses the need for information and the role it plays for an individual as well as for society in its attempt to determine public policy.

An important part of the Statistical Thinking course consists of writing papers. The first paper deals with some aspect of statistics, based on what has been discussed in the course up to that point. A popular topic consists of taking an issue of a magazine like *Newsweek* and discussing the quality of some of the graphs found in that issue. Other topics include the role of statistics in some particular field. The second paper deals with the study of the relationship between two variables; how strong the relationship is, whether it is statistically significant, and whether it is causal or not.

The hope is that the course gives an overview of the role of statistics without getting bogged down in technical details. Some students have taken this course first and gone on to the Statistical Methods course or a mathematical statistics course. They report that they find the second course particularly interesting because they already have some sense of the underlying issues, and they see the need for the particular statistical methods in a new and informative light.

4. Project NABS: Nuts and Bolts Study (Dick Gunst)

This classroom experiment is an introduction to statistical problem solving. It encompasses concepts of quality improvement, the need for data (information), variability, group/team solutions, experimental design, graphical displays, and the scientific method. It requires approximately 30 minutes but can take up an entire 1 1/2 hour lecture depending on how much the instructor wishes to discuss the individual issues raised by the experiment.

Requirements.
1. At least four identical jars are used for the first phase of the experiment. The jars are approximately 7 inches tall and are not wide enough for an adult's hand to fit through the mouths. Each jar is filled with nuts, bolts, and washers. Most of the nuts in the jars fit the bolts, but there is a wide assortment of sizes in each of the jars and there are some nuts that do not fit any of the bolts. Refer to these jars as the "mixed" jars.
2. At least four stop watches, the number equal to the number of mixed jars.
3. At least four additional jars, the number again equal to the number of mixed jars. Half of the jars are identical to the mixed jars. The other half of the jars are smaller, with mouths the same size as the tall ones. Approximately half of each size jars are filled with an assortment of large bolts and their corresponding nuts. The remaining jars of each size are filled with an assortment of small bolts and their corresponding nuts. No washers are included in any of these jars and each nut fits a bolt. Refer to these jars as the "sorted" jars.

Setting. The class members are employees of or consultants to a company that manufactures nuts, bolts, and washers. One very large group of its customers requires that the bolts it purchases be shipped with the corresponding nuts securely fastened. In recent years the company has been losing customers because its competitors have been able to satisfy these customer requirements at a lower cost. Management has mandated that costs be reduced.

Costs of the raw materials and machinery to manufacture the nuts and bolts are standard industry-wide. A cost analysis of the process indicates that the only manufacturing component in which the company's competitors could be achieving lower costs is in the assembly process, the fastening of the nuts to the bolts. In particular, the amount of time needed to fasten the nuts to the bolts has been determined to be the critical factor in the cost differential between this company's costs and those of its competitors. The employees (consultants) have been assigned to a task force to determine the most effective means of reducing the assembly times.

Assembly Process. At this juncture the students are unable to recommend any cost saving measures because nothing is known about the actual assembly process. Any recommendations made at this point are easily recognized as pure conjecture. A reasonable suggestion (made by the instructor) is to familiarize the task force with the assembly process and to collect some data on assembly times.

Industry-wide, nuts are fastened to the bolts by hand. Automation of this procedure has not been found to be cost effective because the nuts and bolts arrive at the assembly stations in large bins containing a wide assortment of sizes of nuts, bolts, and washers. In order to reduce accidents due to spillage, the nuts and bolts are extracted from the bins and placed into jars. These jars are delivered to the individual workers. The workers remove nuts and bolts from the jars and fasten them together, placing the assembled nuts and bolts into sorted bins in preparation for packaging. Data are needed on the time needed to fasten the nuts to the bolts. Stopwatches are deemed suitable for this purpose.

The students are divided into groups in any convenient way. There are at least two students in each group, with a maximum of about six per group. There are a sufficient number of mixed jars and stopwatches to distribute one of each to each group. The task group members are permitted to familiarize themselves with the operation of the stopwatches. They may also examine the jars of nuts and bolts. As they are doing so, they are informed of the specifics of the assembly procedure. Because of safety reasons, OSHA (The U.S. Occupational Safety and Health Administration) mandates the following procedure:

(a) The nuts and bolts may not be poured from the jars (danger of spillage and subsequent employee accidents). Some portion of the bottom of the jar must always touch the table. The jars may be tilted if the bottom remains touching the table.

(b) Nuts and bolts must be extracted from the jar one at a time. More than one nut or one bolt may be placed on the table or desk and remain out of the jar as matches are sought.

Initial Data Collection and Analysis. One individual in each group is selected as the assembler. Another person is the timer. A third individual is the data recorder for the group. Timing begins as the lid of the jar is being unfastened. It stops when four nuts have been completely fastened to four bolts. Once the time has been recorded, the procedure is repeated with different students acting as the assembler, timer, and recorder. No controls (e.g., the same timekeeper) are allowed so that everyone can participate and so the uncontrollable variability due to different assemblers and timers will be evident in the results. If there are at least four groups, it is not necessary to repeat the experiment further.

One of the main features of the analysis of data from this experiment is its simplicity. A point stressed is that often only simple data analyses are required to extract important information from experimental results. This is especially true if the data are collected in a well-designed experiment that includes satisfactory controls over extraneous variability.

As the data are reported, they are graphed using a simple point plot (dot plot) of the assembly times. This is done on an overhead projector or on a blackboard. Usually the times range from about 50 to 200 seconds, although occasionally an observation is higher than 200 seconds.

The data are usually widely scattered. Comments are made about how the experiment was conducted under controlled conditions (one classroom, same time of day, same instructions and training, etc.) and yet the data vary widely. The concept of variability is stressed in this discussion. Students are alerted to the potential difficulty in interpreting data that exhibit such variability, motivating their presence in the course.

Causes/Solutions. Next students are asked to critique the assembly process, with special attention to improvements that would reduce the assembly times. With very little encouragement, numerous ideas are suggested, occasionally peppered with a sense of frustration! In this "brainstorming" session, all ideas are sought and are written on an overhead transparency or on the blackboard. Time permitting, these ideas are organized (e.g., cause and effect diagram) and prioritized.

In virtually every session, three of the ideas for improvement are:

- sort the sizes of the nuts and bolts,
- select a different size for the jars, and
- eliminate the washers.

While all the ideas are praised, these three are selected as especially noteworthy. The remainder (e.g., can't pour nuts and bolts from the jars, training, time of day, differences in stopwatches, differences in assemblers and timers, heights of desks, etc.) are discussed in terms of continuous quality improvement and the scientific method, with the end result that some will be selected in further rounds of experimentation.

Primary Experiment. The primary experiment consists of a factorial experiment in two factors: jar size and nut/bolt size. Four or more additional jars are introduced, as described above in Requirement 3. The jars are randomly distributed to the groups. Prior to this phase of the data collection, students are asked their expectations. Most state that the times will be reduced with the smaller jars. Some also believe the larger nuts and bolts will reduce assembly times.

As times are reported, they are again plotted. A single point plot is constructed beneath the point plot for the initial data. On the second plot, data from each of the four combinations of the two factors are plotted using different symbols. Two things typically occur. First, the four plotting symbols overlap, revealing that no combination is clearly preferable to any of the others. Second, contrary to any of the student conjectures, *all* the plots show reduced assembly times; i.e., there is a downward shift in the second set of plots from the initial assembly times plot.

Conclusions. So long as some sorting is performed and the washers are eliminated, the assembly times will typically be reduced. The company can accrue additional cost savings by selecting the jar size that minimizes cost. These benefits are the result of simple experimentation and simple data analysis, both of which are based on sound statistical principles.

Appendix C: Making It Happen

Interesting and Available Data (Robin Lock, St. Lawrence University)

The purpose of this column [in the *Statistical Computing and Statistical Graphics News-letter*] is to provide a means for sharing the sorts of interesting data that stimulate, challenge, and enliven the practice of statistics. ... Contributed data sets will be made available through the StatLib file server at Carnegie Mellon University. For example, to obtain the baseball data send a message which contains only the line

```
send baseball.data from datasets
```

to the address

```
statlib@lib.stat.cmu.edu.
```

A list of all data sets currently available can be obtained by sending the message

```
send index from datasets
```

to the same address. General information about StatLib is available by sending the message

```
send index
```

to the StatLib address. (From *Statistical Computing and Statistical Graphics Newsletter*, April 1991.)

EDSTAT-L: Statistics Education Discussion List (Tim Arnold, North Carolina State)

Announcing a new electronic forum for discussion related to statistics education: the purpose of this list is to provide a forum for comments, techniques, and philosophies of teaching statistics. The primary focus is that of college level statistics education, both undergraduate and graduate studies.

This list exists because we can learn from each other. What techniques are we using in statistical instruction? What strategies should we use to prepare our students for the future? What part do or should computers play in instruction? Are computers a new tool or a new subject for students? What assets (e.g., public-domain programs, data sets) can we share among ourselves? There are many individuals working in these areas, each with knowledge valuable to the others. This list attempts to bring together every teacher, student, researcher, and specialist interested in improving statistical instruction.

You are encouraged to:
- Submit summaries of or commentary on published articles and books which address problems or solutions in teaching statistics.
- Share the results of your experiments in teaching statistics.
- Ask thought-provoking questions about the future of statistics education and the environments in which it takes place.
- Recruit interested parties to this list, regardless of their status in education, industry, or government.

Subscribing to EDSTAT-L

Bitnet Users: Send the following command as the first line of a mail file to `list-serv@ncsuvm`:

> `subscribe EDSTAT-L Your full name`

IBM VM users in Bitnet may add themselves to the mailing list with this command:

> `tell listserv at ncsuvm subscribe EDSTAT-L Your full name`

VAX/VMS users in Bitnet can subscribe in a similar way:

> `send listserv@ncsuvm subscribe EDSTAT-L Your full name`

Internet Users: Send the following command as the first line of a mail file to `list-serv@ncsuvm.cc.ncsu.edu`:

> `subscribe EDSTAT-L Your full name`

Appendix D: Report of Workshop on Statistical Education

by Robert V. Hogg, University of Iowa

On June 18-29, 1990 thirty-nine statisticians gathered for a workshop on statistical education at the University of Iowa in Iowa City, Iowa. These persons represented universities, colleges, consulting firms, business, and industry. This workshop was sponsored by the University of Iowa and the American Statistical Association (ASA). Financial assistance was provided by the National Science Foundation, the ALCOA Foundation, the Ott Foundation, the Statistics Division of the American Society of Quality Control, and the Quality and Productivity Section of ASA.

As we prepared for the workshop, most of the participants, and several others not attending, wrote position papers on some aspect of statistical education, the majority of which concerned a first course in statistics. As a group we recognized several poor characteristics of science and mathematical education, including statistical education:

- Improvements are needed in the K–12 curriculum as there is widespread science and mathematics illiteracy in the United States.
- There is little effort given to recruiting students to these areas, often because we want only students like ourselves.
- There is a lack of truly qualified teachers because the pool from which they come is drying up.
- College and university instructors are often required to teach large introductory classes that allow little or no interaction with students and severely limit student involvement with others.
- Frequently students view courses as being "hard" because the class periods are dull and it is difficult to get good grades despite spending long hours doing homework.

Reprint of a report of a workshop on statistics education, especially at the introductory level, organized by Roberg Hogg and held June 18-20, 1990.

- Sometimes there is a communication barrier. The instructors often forget to emphasize the "big picture" by concentrating on the solutions of the problems under consideration and eliminating any in-depth discussions of the basic ideas.
- In most universities and some colleges, there is no sense of community among most students in these areas. Accordingly, there is not much discussion during and after class periods. Somehow we should encourage teamwork and stress the importance of it.

Many of us in higher education use the excuse that the students should be better prepared. They probably should be, and there are many efforts to improve education in K–12. However, the fact that we have these poorly prepared students does not excuse us from doing much better than we do. Besides, many of these "poor" students are really extremely good, and we, in colleges and universities, are turning them off because:

- We fail to communicate our enthusiasm and excitement about mathematics and science (in particular, statistics).
- We do not encourage teamwork, which often results in the joy of discovery through lively discussion, observation, and analysis.
- There are "grade wars" and students believe that they must beat others to get high grades.
- Formulas are often not motivated and are presented as an "end" rather than the "means."
- Introductory courses have low priority among faculty and administrators, and students observe this.
- Computer use is perfunctory or nonexistent.

Many of us believe that those of us in universities must "reach out" more to a number of groups, including:

- The high schools, in particular working with future and present high school teachers, strengthening mathematics and science requirements and curricula, and often serving as role models.
- Colleges, with much more interaction between university and college professors and students.
- Graduate students, working with them on their teaching as well as their research.
- Junior faculty, serving as mentors in teaching as well as research.
- Colleagues in other fields, interacting and working on joint research projects.
- Business/industry/government, because many more joint efforts are needed here if America is going to continue to be in the "first class" category. Moreover, these persons can offer faculty and students glimpses of practical applications.
- The public, trying to improve numeracy and science literacy.

However, it is time-consuming to do these things. Thus we must change the present system if our faculties are to get credit for these efforts. As it is now, most of us find it to our advantage to be "loners" doing research. None of us are against research; we support it strongly. On the other hand, we seek a better balance in our reward system recognizing that good teaching and valuable service are as important as research. We want to recruit strong, but possibly different, students to science and mathematics by improving our courses.

The preceding comments apply to statistical education as well as science and mathematics in general. We believe that some of the leaders in our profession must help in serious efforts to address these concerns, and we hope this report will encourage them to do so.

Certainly ASA has taken some very positive steps with its Quantitative Literacy program in improving statistical presentations in K–12. Much more can be done, and it will take enormous efforts on the part of many to shake up the system enough to change science and mathematics education for the better.

It is interesting to note that the "second Sputnik" (Japanese cars, electronics, etc.) has arrived for industry and business, but an equivalent impact on science education as followed Sputnik in the late 1950s and 1960s has not occurred. Some of us had hoped that the 1983 report *A Nation at Risk* would shock our leaders into action. Even though it was short and printed in big type, we are afraid few leaders read it—or, if they did, they did not take it seriously enough to cause substantial action. Thus, we continue our decline to becoming a second-rate nation. Wake up America! Let us create demand for a return to quality education which is competitive with other industrialized countries.

Problems in Statistical Education

Clearly in a three-day workshop we could not address all of the problems listed in the preceding section even though they apply to statistics as well as to science and mathematics. We thought that we could do the most good by addressing the problems associated with our introductory courses. Inspiring courses early in students' college career would help improve their numeracy and possibly encourage some to consider careers in science. We believe students should appreciate how statistics is used in the endless cycle associated with the scientific method: we observe Nature and ask questions, we collect data that shed light on these questions, we analyze the data and compare them to what we previously thought, new questions are often raised, and on and on. It has proved to be an exciting process for many years; but, most of the time, the students do not observe this excitement in our classes. There is a consensus among statisticians that statistics offers a unique and useful way of looking at the world and helping to solve real problems—or to exploit opportunities, but that statistics teaching in colleges and universities fails to communicate the potential, the utility, or the joy of discovery through investigation.

Unfortunately, statistics courses are seldom designed with any idea of what it is that students are supposed to be able to do as a result of having taken the course. Course objectives, if formulated at all, are expressed in terms of the specific topics covered in elementary texts rather than meeting the needs of the students. Statistics is seen as a "subject" rather than as a problem solving tool to be used in the scientific method, or as a useful way to look at the world around us. Some of the many problems associated with our introductory courses include:

1. Statistics teaching is often stagnant; statistics teachers resist change. The most popular elementary texts have evolved slowly over the decades. There is a tendency to present the same subjects, the same way, from the same books year after year. Meanwhile statistics is progressing rapidly.

2. Techniques are often taught in isolation, with inadequate motivation and with no connection to the philosophy that connects them to real events; students often fail to see the personal relevance of statistics because interesting and relevant applications are rare in many statistics courses. The applications, if any, are often contrived, even "phony." Elementary courses are often taught by the youngest and least experienced teachers, who have limited, if any, personal contact with serious applications. The open-ended

nature of statistical investigations and the sequential nature of statistical inquiry are not brought out. The students are not pushed to question their environment and seek answers through investigations.

3. Statistics is too often presented as a branch of mathematics, and good statistics is often equated with mathematical rigor or purity, rather than with careful thinking. ("Statistics has no reason for existence except as a catalyst for learning and discovery." — George Box.)

4. Teachers are often unimaginative in their methods of delivery, relying almost exclusively on traditional lecture/discussion. They fail to take into account the different ways in which different students may learn, both individually and in groups, or the many possible modalities of teaching.

5. The "customers" of the statistics course are often the statistics teachers themselves—not their students, not other areas in which the students study, and not the students' ultimate employers. There is little attempt to measure what statistics courses accomplish; statistics is too little *used*.

6. Many teachers have inadequate backgrounds in knowledge of the subject, in experience applying the techniques, and in the ability to communicate in English. The word "statistics" has itself acquired bad connotations.

7. Statisticians may put their subject in a bad light for the students. They often fail to see any need to convey a sense of excitement.

8. Some teachers are technically incompetent, either in aspects of statistics or in the underlying mathematical tools. They may mislead by treating statistical investigations as if they entailed random sampling from some finite population.

9. Statistical notation is unnecessarily complex and inconsistent; the Greek alphabet is routinely used rather than reserved only for those occasions in which it is essential.

Although the problems of statistical teaching are severe, the picture should be seen in proper perspective. Innovative texts are available and achieving some degree of market penetration. More are needed. However, statistics teachers often face obstacles to good teaching that are imposed by the institutions for which they work, such as the following:

a. There are the unpleasant realities in college and university teaching environments, like "mega-classes" which make it hard to do anything beyond lecture/discussion and seem almost to be designed to make students dislike the subject matter and to spur them into apathy. There are less extreme, but still annoying, constraints such as the impossibility of changing short class periods into longer periods thus allowing for laboratory-type instruction and learning.

b. University incentives do not encourage faculty to collect good examples, improve teaching methods, or spend time inspiring students.

c. The content of the courses (including the text) may be specified by others, leaving little freedom for an innovative statistics teacher to make improvements. The course may be required to attempt to cover more material than is reasonably possible.

d. Location of statistics courses in the mathematics department is seldom ideal.

e. Traditional grading systems may perform some necessary functions, but they often divert students from the fundamental objectives of learning statistics as well as other subjects. Examination requirements may be unnecessarily rigid.

f. University logistical support for statistics teaching (and teaching in other areas as well) is often weak. Students may have no or, at best, limited access to computers or good interactive statistical software. Classroom amenities like good projection equipment may be lacking.

g. Some obstacles are posed by the students themselves. While the instructors must try to motivate students, it is difficult to change the attitudes of many of them.

h. Students often have a fear of formulas and mathematical reasoning. An even more serious problem is the desire for props, such as rote memorization or problems with quick and clean answers that avoid the need for students to think, but make it easier for them to perform well on examinations. Some students may even like formulas because they are props. Students may be used to reading large volumes of materials relatively superficially, rather than a relatively few pages carefully, as is more suitable in statistical instruction. Unlike some courses, statistics is cumulative and does not lend itself to crash-cramming sessions at the end of the term.

i. Statistical thinking is different from that to which most students have been exposed. The ideas of uncertainty and variability are either ignored or dealt with poorly in everyday discourse and even academic study. Statistical thinking may even be counter-intuitive, as illustrated by the appearance of "hot streaks" and "momentum" or the notion that the "sophomore jinx" is a mysterious plague of nature.

j. Much of the statistical reasoning students encounter in everyday life represents misuse rather than sound use of statistics. The bad reputation of statistics courses may become a self-fulfilling prophecy.

General Suggestions

During the workshop participants were divided into four teams. Most of the time was spent in team meetings, with occasional plenary sessions to summarize the activities of each team. Although the four teams met separately and had lively discussions, there were several common suggestions.

STATE OUR GOAL(S)

Stating the aim of the course is an important, and often over-looked, step toward successful teaching. Participants spent much time debating the skills and knowledge they expected to impart to students in an introductory course, as opposed to the list of methods they wished to cover. One team came up with the following:

> Our aim in a first course is to develop critical reasoning skills necessary to understand our quantitative world. The focus of the course is the process of learning how to ask appropriate questions, how to collect data effectively, how to summarize and interpret that information, and how to understand the limitations of statistical inferences. Statistical thinking is central to education.

Unfortunately, the typical introductory statistics course does not meet this goal, as it stresses mathematically precise statements and formulas applied to artificial data that are of little, if any, interest to most students. A typical textbook example begins with a question that has been formed to address one feature of data that has already been collected. Students gain little insight into how and why data are collected, how experiments are designed, and how analysis of one set of data leads to new questions, new experiments, and subsequent analyses in a continuing cycle of scientific inquiry. Statistics is presented as a formal ritual, rather than as a dynamic study of processes.

Every team agreed that the typical introductory course should change by giving more attention to graphical techniques, to simple topics in the design of experiments, and to the scientific method, with much less time to hypothesis testing. Most introductory textbooks devote a great deal of space to hypothesis testing, which workshop participants saw as being much less important than current coverage would indicate.

There was some agreement that this type of statistics course is valuable for students of all interests, although some participants favored the idea of tailoring courses to different clienteles. Another goal could be the identification of the interests of a group of customers and focus the course content and presentation on the needs of that group, possibly through joint efforts of statisticians and members of the other group.

There was also agreement that undergraduates interested in mathematical statistics, who are the largest pool of future graduate students in U.S. statistics programs, should depart from current standard practice and also take an additional introductory course in statistics that features data analysis and applications.

ANALYZE DATA AND DO PROJECTS

There was widespread agreement among workshop participants that students should work more with real data and with graphs. Many advocated projects in which students collect their own data and analyze them in written reports. Such projects combat apathy by allowing students to work with data that they find interesting. Moreover, projects give students experience in asking questions, defining problems, formulating hypotheses and operational definitions, designing experiments and surveys, collecting data and dealing with measurement error, summarizing data, analyzing data, communicating findings, and planning "follow-up" experiments that are suggested by their findings.

TO COMPUTE OR NOT TO COMPUTE

Properly used, the computer is a powerful and effective tool in teaching students about variability in data, particularly through statistical graphics. Most participants strongly support the use of the computer in introductory statistics courses. Having a computer available during class facilitates "on the spot" analyses of data, which teach students that there are often many analyses that can shed light on a problem and that simple graphics can tell one a great deal.

However, a variety of problems, including lack of equipment, lack of adequate projection systems, and lack of staff to operate statistical computing laboratories, severely limit current computer use in statistics courses. Indeed, a small group of participants saw such obstacles, coupled with the concern that teaching students to use the computer can shift attention away from statistical ideas, as sufficient reason to avoid computers in their teaching at the introductory level.

LECTURE LESS, TEACH MORE

Much time was spent in team sessions exploring various modes of teaching and learning. Lecturing is but one way to communicate ideas to students. Classroom demonstrations and data collection (for example, analyzing the color distribution in a bag of M&Ms, or examining the distribution of the numbers of M&Ms in several bags), add variety, as do experiments that are run in class. Much effective learning takes place when students work together in teams. Student projects can be done in small groups, but some experiments can be conducted in class.

Of course, some teachers are faced with very large classes. Two ways to make a large class seem small are emphasizing project work and lecturing to a small sample of students who are brought to the front of the room at the beginning of each class period.

WHAT CAN THE PROFESSION DO TO HELP?

It was repeatedly suggested that a "First Course Corner" or "Activities Corner" be added to *The American Statistician*, in addition to the existing "Teacher's Corner," which deals mainly with issues that are only of interest in teaching graduate level courses. Such a new section in this publication would provide a place to share interesting sets of data, experience with projects and classroom experiments, and other ideas to improve the introductory course. It would also signify a commitment on the part of the profession to improve the teaching of introductory statistics courses.

Other suggestions included short courses at meetings to teach teachers about such topics as process improvement (which few current statistics teachers studied in graduate school) or the effective use of student projects in the introductory course. Of course, it would be quite helpful if universities assigned their best teachers (and those most experienced with analyzing data) to introductory courses and reward appropriately excellent and innovative teaching.

In addition, some participants of the workshop are developing sample syllabi to share with other teachers. This idea has been used successfully in the calculus reform movement in the U.S., as it helps the hesitant teacher pare old lists of topics and methods down to the essential items, thus creating space in the curriculum for fresh ideas, such as student projects.

Detailed Suggestions

Clearly, with 39 individuals grouped into four teams there were many suggestions and there certainly was not unanimous agreement on each of these. However, listed below are suggestions that most of the participants supported. The first list consists of topics that seem important for a first course (1-4 are highest priority, 5-9 second, 10-13 third, 14-17 fourth).

IMPORTANT TOPICS FOR AN INTRODUCTORY COURSE

1. Recognizing that statistics surround us in everyday living. Reported statistics are sometimes incorrect or misused; thus it is important for each of us to be a critical consumer of statistics given by the media. We must ask questions about the quality of the data and the reliability of the analysis.
2. Understanding variability: bias, sampling error, systematic error, measurement error, regression effect, etc. In particular, understanding "Actual Observation = Fitted + Residual," and that in statistics we try to detect the pattern (fitted) and describe the variation (residual) from that pattern.
3. Collection and summarization data, including basic exploratory data analysis. (Some felt we should do more with writing up and explaining results.)

4. Graphs, including plotting data taken sequentially (that is, basic time-series concepts).

5. Sampling and surveys, including the importance of getting quality data.

6. Elementary designs of experiments, with some discussion about the ethics of experimentation and the distinction between observational and experimental investigating.

7. Formulation of problems and understanding the importance of operational definitions and the process of inquiry. That is, understanding the iterative nature of scientific method, including: Plan–Do–Check–Act. We want the capability to make and understand predictions.

8. Basic distributions (normal, binomial, etc.) as approximations to variability in data sets: modelling.

9. Correlation and regression and other measures of association. For example, there should be some illustrations of Simpson's paradox.

10. Elementary probability, including event trees and conditional probability (some included Bayes' Theorem).

11. Central limit theorem and law of averages.

12. Elementary inference from samples, recognizing there are not unique answers in statistics.

13. Ability to use at least one statistical software package.

14. Outliers and how statistical measures change with various changes in the data (that is, aspects of robustness).

15. Statistical significance vs. practical significance.

16. Categorical data and contingency tables.

17. Simulation.

In addition to these topics, there were these other suggestions that could be used by instructors and students:

1. We need to produce syllabi so that less confident teachers can mimic more experienced teachers. Nondeterministic thinking often seems unnatural and we must help the inexperienced teacher and student.

2. Let us think more about processes rather than bowl models. Multiple sets of data should be analyzed, perhaps by multiple analyses. Certainly we should understand the difference in what Deming calls enumerative (bowl) and analytic studies.

3. Students rely too much on formulas rather than on thinking. Do everything possible to improve student's self-esteem and his or her appreciation of and self-confidence in statistical studies.

4. Even with large classes, we need some interaction with students. Try using a small subsection as "your class" with the others as observers. Also select a few students each day for Mosteller's one-minute drill (list on a small sheet of paper the most important topic of the day, the muddiest, and the one about which the student would like to know more—or other short appropriate comments).

5. Find appropriate articles, in the popular press and technical sources, for students to read. Think of projects (questions) for which students can collect and analyze their own data.

6. Statisticians are sometimes overconfident, even arrogant, about the intrinsic importance of their subject. This puts statistics in a bad light, and we should be aware of this and

try to change this attitude.

7. Occasionally have a guest lecturer or colleague in another field or one from industry or government (could consider retired statisticians). These might be held as informal sessions in the late afternoon or evening. The setting should encourage interactions between students and the visitor so as to expose the skills of an experienced statistician.

8. Work on improving our presentations (there was one suggestion that many of us should take acting lessons). Some find it valuable to include some humor into courses.

9. Provide extra problems and old exams through the many "copy stores" that most colleges and universities attract.

SUGGESTIONS FOR THE PROFESSION

Other suggestions were made that cannot be accomplished by an individual instructor, but we would need an effort by a collection of people, like the ASA:

- Constructing a journal on teaching statistics like the MAA's *The College Mathematics Journal* (at least have a "First Course Corner" in the *American Statistician*).
- Newsletter on statistical education.
- Providing information on teaching TAs to teach (MAA has one which was edited by Bettye Anne Case). Encourage extensive TA training and mentoring.
- Developing teachers' network.
- Workshops on teaching at ASA meetings.
- Short courses on teaching.
- Poster sessions on teaching, student projects, and data sets.
- Supporting faculty efforts to change the system.
- Diagnostic tests for students.
- Encouraging position papers, each prepared by two or three like-minded persons.
- Funding for future conferences or workshops on statistical education.
- Collection of "tidbits," namely good ideas for classroom teaching.
- Writing a "Point of View" article for last page of *The Chronicle of Higher Education*.
- Supporting efforts to modify the academic system, in particular, the reward structure and grading procedure.
- Becoming aware of difference of jargon. For example, an engineer might call the statistician's independent variable (or factor) a "parameter."
- Center for collecting materials like projects, videos, and case studies.

Final Thoughts

Clearly, with 39 statisticians involved, we could not agree on the topics to be included in that first course. There will be a continuing need for several first courses. For example, one person clearly thought χ-square tests must be in such a course because his colleagues at his college expected it (and he thought his students appreciated it). Most, however, would decrease the role of tests of hypotheses; some eliminated that topic altogether.

We should ask ourselves what it is that we want the students to remember about that one statistics course that they took ten years ago. We hope that it is not "The worst course I took in college," although that is frequently the answer that we get today. In thinking about ways to improve the answer, we should re-read the goals of the course as suggested above. Whether we agree with this or not, it will make us think hard about the process of continually trying to improve our introductory course.

We certainly have not solved everything and we must continue our efforts to improve statistical education. Many suggestions have been made in this report on how to do so. Certainly we should seek appropriate funding to get statisticians together to produce tangible materials. If we have future workshops, each team probably should focus on a very special and important topic (at this one, all teams worked on "the first course").

At professional meetings, we must have sessions devoted to statistical education, particularly what improvements are actually being made in these early courses. However, one good sign is that ASA is doing something: in addition to the QL program and the Center for Statistical Education, the topic of the 1992 Winter Meeting is Statistical Education. Hogg has agreed to serve as program chair, and a huge effort will be made to get graduate students to attend. One way to do this is to get many employers (academic, business, industry, government) there to interview students. This Winter Meeting is an excellent time for these interviews. But, beside these employment opportunities, we want to provide a program that students and those interested in statistical education can enjoy. This includes statisticians from industry, business, and government too, because much of the program will be devoted to education beyond that in the traditional academic community.

So, as we left our workshop, there was a great deal of determination that we would not let this topic drop. Education is too important to our profession.

Special Thanks

As indicated in the introduction, we appreciate the sponsorship of the University of Iowa and ASA, as well as those groups that helped financially. In addition, there were the leaders of the four teams: Harry Roberts, Bob Miller, Tom Moore, and Ed Rothman/Bobby Mee. Ed and Bobby jointly took over the fourth team at the last minute. There were some excellent note-takers in Mel Alexander, Jim Sconing, and Jim Calvin. Three persons stayed an extra day to help organize all the notes and thoughts, namely Harry Roberts, Jeff Witmer, and Peter Hackl. Hogg, working under considerable strain due to personal reasons, would have had a most difficult time preparing this report without their help.

Finally, it is fair to say that all participants took this workshop very seriously. While there were disagreements (a few wanted more revolutionary actions suggested), every one of the 39 statisticians made some contributions and their work before, during, and after the meeting is truly appreciated.

Workshop Participants

Mel Alexander, Jim Calvin, Patti Collings, Jon Cryer, Marilyn Dueker, Andrew Ehrenberg, Herman Friedman, Barry Griffen, Bert Gunter, Peter Hackl, Gudmund Iversen, Mark Johnson, Brian Joiner, Jim Landwehr, Bob Lochner, Richard Madsen, Bobby Mee, Bob Miller, Tom Moore, Peter Nelson, Ron Patterson, Harry Roberts, Tim Robertson, Ed Rothman, Ed Schilling, Jim Sconing, Elliot Tanis, Aaron Tennebein, Neil Ullman, Joe Voelkel, Ray Waller, Ann Watkins, Roy Welsch, Carl Wetzstein, B.F. Wink Winkel, Jeff Witmer, Liann Yuh, and Arnold Zellner.

Tomorrow's Geometry

Tomorrow's Geometry

Joseph Malkevitch
York College (CUNY)

Introduction

In *A Century of Mathematics in America (Part II)*, Robert Osserman contributed an article entitled "The Geometry Renaissance in America: 1938-1988." The renaissance in geometry that he recounts has not been restricted only to America and did not end in 1988. Nor, as Osserman notes in a "postscript" to the article, has this renaissance been confined only to the area of differential geometry, which is what his article deals with. There are many reasons for the renaissance in geometry, ranging from new developments in physics and biology, to the development of the digital computer, to the flowering of old and new fields within mathematics that stimulate geometric insights.

Yet in terms of the way geometry is represented in the undergraduate curriculum, there has been no renaissance. On the contrary, there has been relatively little change in the nature of the geometric mathematics taught in colleges. This unchanging curriculum for geometry in college has in turn prevented any new geometry or applications stemming from geometry from being presented in high school (or earlier grades). Students planning to become high school or middle-school teachers usually see only that geometry covered in a survey course on Euclidean and non-Euclidean geometry. The content of such courses rarely discusses geometry that reaches into the period of which Osserman speaks. However, because of its unique intuitive and quick-starting nature, geometry that has been discovered in the last 30 years can be taught to undergraduates.

In light of these contrasting phenomena—explosive growth in geometric methods and applications of geometry but an unchanging undergraduate geometry curriculum—the MAA Curriculum Action Project convened an E-mail Focus Group to debate geometry education. A diverse group of research geometers and geometry educators agreed to participate in this electronic debate.

Questions To Grow On

In order to help participants in the electronic mail discussion think about the issues, key questions were formulated as a framework within which an informed exchange of views on educational reform for geometry could take place. In this introduction, each of the questions is followed by discussion that motivates the choice of the question as one of those to frame the debate. Although this background information appears here, only the questions were posed prior to the original electronic mail debate.

Question 1: What steps do you feel should be taken to improve the quality of undergraduate education in geometry? For example, what new courses might be developed? What changes are needed in existing courses? Are there suitable ways to include geometric topics in courses not specifically about geometry?

Background: Most undergraduate courses dealing with geometry belong to one of two types. One type of course, often designed for students bound for graduate school, is devoted to some specific part of geometric mathematics. Examples of such courses include Differential Geometry, Convexity and Geometric Inequalities, Theory of Graphs, Topology, Combinatorial Geometry, or Knot Theory. These courses (especially at smaller colleges) have often been developed by a member of the faculty with a research interest in the area, and are often taught exclusively by one faculty member. Such courses sometimes fall into disuse when this faculty member moves on to another college (or retires).

The other type of course will be referred to, with perhaps a slight abuse of language, as a "survey" course. These courses are sometimes taken by students bound for graduate school, but the most common type of student in such classes are prospective high school teachers. These courses are taught under such titles as Advanced Euclidean Geometry, Modern Geometry, Non-Euclidean Geometry, Projective Geometry, etc. Another common course title, often aimed at high school teachers and often having "survey" overtones, is Geometric Transformations.

Although courses of the first type often deal with mathematics developed since World War II, courses of the second type almost never do. The first type of course often provides depth but not too much breadth. The second type of course provides greater breadth, together with some depth, but is of little research interest to contemporary geometers. Both types of courses rarely mention applications of geometry external to mathematics.

Geometry enters the curriculum not only through courses specifically designed for their geometric content, but also in other courses which may either be required or elective for a mathematics major. Examples include geometric reasoning behind the theory of relative maxima and minima (and points of inflection) in calculus, and symmetry of polygons, polyhedra, and tilings when studying group theory in a modern algebra course. Yet it is not clear that conscious efforts are being made by textbook authors and course designers to systematically exploit the power of geometric reasoning and phenomena in non-geometry courses. In particular, geometry is so under-represented in terms of content and methodology that typical practitioners of mathematics may lack sufficient background to draw on a reservoir of geometric materials for their teaching.

Question 2: What strategies can be employed to get more modern geometric topics into the undergraduate curriculum, particularly for students who may only take introductory level courses?

Background: Most students who take courses designed for mathematics majors in their freshman and sophomore years do not in fact major in mathematics. Yet such students often enter fields in which mathematics plays a significant role (e.g., physics, engineering, and computer science). More often than not, these students take the calculus sequence and linear algebra but are unlikely to take a course with a high density of geometry.

As recently as the late 1950s and early 1960s it was not uncommon for students to precede their taking of calculus with a course in Analytic Geometry. Not only is this no longer true, but increasingly less and less attention to geometric ideas is being given within calculus courses. This is unfortunate, since there are many geometric approaches to calculus that afford insights into such fundamental ideas of calculus as limits, the derivative, area, arc length, etc. Traditional analytic geometry is hardly modern in flavor and what little is left of this subject in current calculus courses is rarely forward-looking. One avenue that has been suggested to get more modern geometric ideas into calculus is via applications. For example, the field of robotics is a very active one and includes many questions of a geometric kind. (Examples include geometric questions arising from motion-planning problems, from questions concerning recognizing objects from various types of probes, and from geometric algorithms involved in questions concerning grasp and motion of robot arms and fingers.) Other fields that might prove to be fertile sources of ideas for bringing geometry into calculus are image processing, data compression, geometric modelling, and computer vision.

Linear algebra has, in the hands of some teachers, proved to be a more fertile area to introduce geometry. Geometric transformations, robots, linear programming, computer vision, and graphics are all subjects rich in geometrical content in which linear algebra is intimately involved. Graph theory also has many important interfaces with linear algebra, for example, in the study of Markov chains. Unfortunately, the introduction of geometric material, though it is closer to the surface here than in some other branches of mathematics, is not the norm. In fact, although algebra and geometry are handmaidens, the geometry behind and connected with linear algebra is often ignored.

However, a success story for modern geometric ideas of a specialized kind has been the creation of a number of courses in discrete mathematics for both mathematics majors and computer science majors. These courses often include extensive studies of the theory of graphs, of geometric algorithms (including in some cases topics from the fast growing area of computational geometry), and of cellular automata. Ironically, this course has been perhaps more successful in introducing computer scientists to geometry and geometric thinking since for them it is usually a required course. Some mathematics departments still refuse to grant credit to mathematics majors for discrete mathematics courses. Often discrete mathematics is a freshman course for computer science majors, which means that the geometric ideas they are exposed to becomes an early tool for them to exploit in later courses. For mathematics majors, any exposure they have to geometry is likely to be deductive geometry, taken at a much later point in their training.

Question 3: As a result of the recent recommendations regarding the preparation of teachers of school mathematics, what specific preparation in geometry is needed by prospective and practicing teachers?

Background: High school teacher preparation and the up-grading of geometric skills and knowledge among currently practicing high school teachers raises some especially difficult problems. Historically, the high school curriculum in the 10th grade has concentrated on the teaching of deductive Euclidean geometry with greatly varying degrees of rigor. So-called two-column proofs of Euclidean theorems have often been the goal in "elite" high schools. However, over the last ten years there has been a growing interest in teaching geometry at least in part in an inductive/problem-posing environment with support from educational technology. There are indications that many high school teachers teach Euclidean geometry as the geometry of the physical universe, and many teachers are unaware of the 19th-century discovery of non-Euclidean (Lobachevsky) geometry.

To give future teachers tools for teaching the existing curriculum, many colleges require high school teachers to take what is referred to as a "survey course" in geometry. These courses aim to show prospective high school teachers a unified framework for some of the results taught in the traditional high school course (e.g., Ceva's theorem as a way to understand the concurrence of the medians, altitudes, and angle bisectors in a triangle) and to introduce them to the fact that other geometries than Euclid's can be consistent. Little time is left over to show future high school teachers such topics as tilings, polyhedra, box packings, etc., which are Euclidean in framework but much more modern in flavor than the circle, triangle, and quadrilateral geometry currently emphasized to the exclusion of nearly every other aspect of geometry.

Furthermore, few high school teachers are prepared, as a consequence of the geometry they learned, to alert high school students who show early interest in research in mathematics to anything other than a distorted view of the nature of research in geometry. Westinghouse projects and mathematics fair papers too often deal with the geometry of triangles, circles, and quadrilaterals which no longer have virtually any interest to geometers. Yet such quick-starting geometric areas as tilings of the plane, graph theory, polyhedra, and convexity are denied an army of bright assistant investigators. The Pascal triangle still often reigns supreme where, say, the study of polyominoes offers an equally rich array of lessons for young students. If teachers are not exposed to these concepts, they will be unable to pass such ideas along to their students.

Question 4: What role can computers and other new technologies play in improving the quality of undergraduate education in geometry?

Background: Over the last few years a variety of computer graphics tools, geometric tool kits, and computer languages with geometric primitives have been developed. Examples include large symbolic manipulation packages (such as *Maple, Derive*, and *Mathematica*) with elaborate two- and three-dimensional graphing techniques; specialized graphing packages designed to graph functions and surfaces; geometric tool kits for Euclidean geometry and geometric transformations (such as the "Geometric Supposer" and the "Geometer's Sketchpad"); computer environments to study graph theory (such as "Netpad," developed by Bellcore); CAD/CAM systems; and the computer language *Logo*.

All of these tools have profound implications for both research and teaching in geometry. These educational technologies make it possible to display and conduct calculations on geometric objects that would either be impossible or very time consuming with paper-and-pencil tools. Although some schools have been experimenting with using these technological tools in lower grades (e.g., *Logo* is introduced in some school systems as early as the fourth grade) and in high school (e.g., the "Geometric Supposer" and the "Geometer's Sketchpad"), most of these tools have remained largely experimental, used in industry but having little effect in classrooms. Other new technologies with research or pedagogical implications are videotape and videodisk materials with a geometrical theme and the availability of supercomputers to solve calculation-intensive geometric problems.

One particular problem with new technologies has been that their evolution has been so fast that college educators are often reluctant to promote any particular one of these software packages, languages, or environments for fear that what they are teaching will all too quickly become obsolete or outmoded. Yet many of these tools have grown and improved as they have matured. A good example is perhaps provided by the language *Logo*. Pioneered at MIT as a means for introducing elementary school children to geometric concepts and exploratory environments, this language quickly became mired in innumerable different dialects (even for the same computer), thereby limiting the tremendous opportunities implicit in this marvelous tool. Geometers (who must share some of the blame here) have been very slow in showing how to use *Logo* in a creative manner (e.g., *Logo* can do more than draw pictures quickly of families of polygons). This has been in part due to fear that the next generation of tools would supersede the current one.

An example of how technology has interfaced with both pedagogical and research aspects of geometry is afforded by the role of computer graphics in the explosion of interest in fractals and dynamical systems. Without computers, the intriguing numerical and visual patterns that have stimulated large amounts of interest among students (and the lay public), not to mention an exponentially growing literature on chaos and related phenomena, would not have been possible.

Question 5: How can more interest in geometry be generated among undergraduate majors in the mathematical sciences?

Background: The traditional first year graduate courses in mathematics are Analysis, Algebra, and Topology. (Introductory topology is rarely taught primarily for its geometric content, but more often as a handmaiden for analysis.) The structure of qualifier and preliminary examinations in universities encourages students to select one of these fields in which to write a thesis rather than such smaller fields as Applied Mathematics, Logic, Number Theory, or Geometry. The result is that relatively few mathematicians refer to themselves as "geometers." As a consequence, when curricular changes are made, geometry's "interests" are rarely protected by persons steeped in either geometry or its methods.

Since most college faculty today were taught a rigorous theorem/proof geometry course in high school, and if they took geometry in college it was likely to have been axiomatically developed projective, Euclidean, or non-Euclidean geometry, it is easy for non-geometers to have the impression that axiomatics are central to current concerns of geometers. Nothing could be further from the truth (see Appendix A). Yet a great gulf exists between the work that geometers currently are doing and the perception, even among practicing mathematicians, about what is currently of interest to geometers. Most current researchers in geometry have little interest in axiomatics but are, rather, interested in developing new questions and methodologies, which, while nominally in the domain of Euclidean geometry, are very far from the traditional concern with the metric properties of circles, triangles, and quadrilaterals.

While Morley's Theorem that the points where the trisector lines of the angles of an arbitrary triangle meet are the vertices of an equilateral triangle (see Figure 1) is a dramatic example of 19th-century geometry (though it belongs to 20th-century geometry!), Chvatal's Theorem that $\lfloor n/3 \rfloor$ point surveillance devices are sometimes necessary and always sufficient to see the boundary of an n-sided polygon (see Figure 2) is a dramatic example of recent elementary geometry. Although Morley's Theorem has many generalizations, none have true research significance for new ideas or methods. By contrast, Chvatal's Theorem inspired a book-length treatment of related problems [Joseph O'Rourke, *Art Gallery Theorems and Algorithms* (Oxford University Press, 1987)], and leads in many directions of interest for mathematics and external applications. These problems are simple enough to explain to middle or high school students and yet generate many new results, methods, and potential application. Fortunately, in addition to the book noted above, other important elementary expositions of recent accessible work in geometry are becoming available.

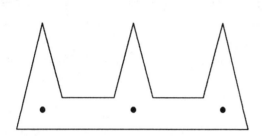

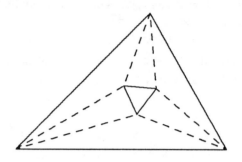

CHVATAL'S THEOREM: $\lfloor \frac{n}{3} \rfloor$ guards are sometimes necessary to "see" the interior of an n-gon.

MORLEY'S THEOREM: The angle trisectors of a triangle determine an equilateral triangle.

Figure 1   **Figure 2**

Another possible avenue for generating excitement about geometry lies in the rapidly growing field of "visualization." The advent of supercomputers has made possible "computation intensive" solutions to geometric questions. More specifically, it is now possible to draw images of complicated dynamical systems. A good example of the interplay of computers and geometry at this level can be found in the book *Computers and the Imagination* by C. Pickover.

Question 6: Should colleges and universities develop an applied-oriented masters degree in the area of geometry? If such degrees were developed, what consequences would there be for the teaching of undergraduate geometry courses?

Background: In August 1989, the Society for Industrial and Applied Mathematics sponsored a conference on Applied Geometry. This conference, held in Albany, NY, was intriguing for the broad nature of the geometric methods and geometric problems that were raised. Some examples include: problems in solid modelling for designing better machine tools, automobiles, and planes; problems in robotics such as local and global motion planning; problems in computer vision such as object recognition; problems in computer graphics such as hidden line removal; problems in image processing such as how to remove blurs from an image; problems in geometric algorithms; problems in computational geometry. Yet what was noteworthy was that relatively few conferees taught in college or university mathematics departments. Most of the people who attended the conference would have described themselves as engineers or computer scientists and, if they were mathematicians, they were working in private industry rather than in university mathematics departments. Although many new methods and ideas were discussed at this conference, many classical methods were alluded to: projective geometry, geometric transformations, barycentric coordinates, splines, polyhedra, Voronoi cells, etc. In some cases it sounded as if the wheel was being re-invented for lack of knowledge of what geometers had accomplished in the past.

Professor Louis Billera (Cornell University) speculated during the conference about whether the circle of ideas that emerged as so useful in these seemingly disparate (but mathematically linked) areas could profit from practitioners with specialized masters degree training in "applied geometry." This question raises many interesting issues.

Historically, careers for undergraduate mathematics majors included high school teaching, actuarial positions, work in industry or government laboratories, work related to computing, operations research positions, college teaching, statistical environments (e.g., Census Department, Labor Department), etc. In order to obtain positions for many of these professions, additional graduate training in mathematics would be necessary, although not necessarily a master's degree in mathematics. For example, a student might go on to get a master's degree in education, mathematics education, statistics, operations research, computer science, decision science, or applied mathematics. The problem, as perceived by many undergraduates, boils down to the fact that there is no section of the want ads of major newspapers which lists ads for mathematicians. In contrast there are sections of want ads for such related fields as engineering and computer specialists.

Many mathematics majors are unclear concerning the value of a mathematics major as a career preparation (especially if they are not interested in becoming teachers at any level). They are also unclear as to what careers a (non-specialized) Master's Degree in Mathematics will enable them to pursue that could not be pursued on the basis of an undergraduate degree alone. If "applied geometry" was a specialty that had clearly identifiable associated career opportunities, the availability of master's degrees in this area would be an additional reason for mathematically inclined students to major in mathematics rather than allied computer, engineering, or scientific fields. Not only might the availability of such degree programs foster greater interest in undergraduate geometry courses, but it also might help forge links between undergraduate and graduate institutions.

Recent Attempts at Geometry Reform

Periodically, geometry makes a great leap forward at the research level, and yet these dramatic changes of direction in research seem to have little effect on the pedagogy of geometry. For example, Joan Richards in her book *Mathematical Visions: The Pursuit of Geometry in Victorian England,* documents the tortuous path by which concepts emerging from discoveries of Lobachevsky, Bolyai, and Klein entered schools at different levels and the mathematics community itself. There are many parallels with the current situation. There have been attempts in the past to lessen the influence of deductive Euclidean and non-Euclidean geometry on the curriculum. Often these reforms have foundered on what is perhaps geometry's greatest strength: its ability to illuminate such a broad collection of phenomena. A typical concern (voiced relatively recently by Edwin Moise) is that at least Euclid is a "theory." If we cut (deductive) Euclid from the curriculum, we will be left with a potpourri of loosely related topics.

In 1967, a conference of geometers was held in Santa Barbara under the auspices of CUPM to discuss future directions for the geometry curriculum. However, it is difficult to assess exactly what effects on curriculum resulted from the conference. More recently, the Consortium for Mathematics and Its Applications (COMAP) held a conference in Boston to study the possibility of changes in the undergraduate survey course and related phenomena. The recommendations from this conference appear in Appendix A. The proceedings of this conference, entitled *Geometry's Future,* also contains an enumeration of various sub-topics in geometry and a brief bibliography for these subtopics. (See J. Malkevitch, "Geometry in Utopia.")

The substantial excerpts from the electronic mail discussions reproduced below reveal a variety of creative ideas to deal with curriculum reform for geometry. However, they also show widespread disagreement about what form these changes should take and how to go about making these changes. We hope that this dialogue will be just the first step towards effective and creative reform.

Dear Geometers:

The quotations that follow are lightly edited excerpts from the electronic discussion undertaken by this Focus Group. They provide various perspectives on the questions raised in the introduction.

I would like to begin with questions about an applied masters degree. I first heard this notion from Lou Billera, though my conception of it and the reason why it appeals as an idea may be very different from his. As we all know you do not find want ads in the *New York Times* under the listing mathematician. Many students are confused as to degree options available. In particular, what jobs are there if one does not get a Ph.D.? At my school, where most of the students are minority, math majors go on to do actuarial work, high school teaching, banking, but rarely go for a Ph.D. Most of our mathematics majors are students who wanted to become engineers but were unable to do well in physics. These students would benefit, as would others who prefer mathematics to engineering, if they could be prepared by a mathematics program for a job. Applied masters degrees in geometry which would combine undergraduate and masters training in computing and mathematics in such areas as computer graphics, robotics, solid modelling, etc., seems to me a way of encouraging students perhaps currently lost to mathematics to stay with mathematics, and study lots of geometry too.

—*Joe Malkevitch, York College, CUNY*

The first wave of thinking about the role of mathematics in computer science always seems to yield "discrete mathematics." But in three important areas of computer science—vision, robotics, and graphics—the answer is "continuous mathematics," especially geometry and geometrically oriented linear algebra and multivariable calculus. I have seen this first-hand in industrial research I have done in robotics and I see it in trying to teach a graphics course to undergraduates at Adelphi.

One consequence is that some computer science programs might like a geometry course for their majors. If there is no room for that, they might appreciate a more geometric flavor for the courses in linear algebra and calculus. (How many linear algebra courses deal with isometries, affine maps, and projective transformations?)

In regard to graduate work, one might suppose that a degree in applied geometry is too narrow to attract students. As an example that this may not be true, I mention that New York Polytechnic offers a masters degree in image processing (that may not be the exact title). Geometry is at least as broad as that. However, I think one needs to take into account that most people go to graduate school as a way of targeting a particular career opportunity. There is a "computer industry" but not a "mathematics industry" in quite the same sense. Perhaps one might devise a degree which had a substantial amount of "hands on" computing, and get some cooperation from computer science faculty. Then if one found a good title and effective marketing (how to do that?) one might make it fly. However, it might be something more risky or ambitious than any individual school would want to try. Is there an NSF program to subsidize ventures like this?

—Walter Meyer, Adelphi University

I have some additional thoughts to share about geometry and its position in the mathematics majors. Different mathematics majors, depending on their career goals have different needs. Historically, it seems to me that overly great concern has been given to undergraduate training in relation to future Ph.D.'s in mathematics. Most undergraduate majors do not go on to get Ph.D.'s. We are at a point where very few students show an interest in studying mathematics (no less geometry) in college. I believe not only can we be doing more to encourage students to major in mathematics, but also to study more mathematics as part of their major.

The first course a student sees, typically, is calculus. This course has had interesting geometry all but squeezed out in recent years. (There was a time when analytic geometry preceded calculus and many geometers enjoyed that, but I do not think we can turn back that clock, nor should we.) However, my point is this: If we are thinking about success for a Ph.D. program and a student appears lack-luster in calculus, many teachers will encourage students to try a different profession. Yet geometry skill and devotion are often exhibited independent of performance and skill in many of the mainstream undergraduate mathematics courses.

We need a way to identify talent and interest in geometry. One of geometry's appeals is that many parts require relatively little prerequisites. The course that Thurston, Doyle, and Conway are teaching at Princeton, open both to mathematics and non-mathematics majors, is intriguing in this regard since it is a serious experience for mathematical types, allows an early experience with geometry, and offers the chance of attracting students who by their exposure to mainstream mathematics may have not even had a chance to see what mathematics, no less what geometry has to offer them.

If we are to include recent geometry in the undergraduate mathematics major, a significant change will have to be made in the way the survey course is taught. In courses in differential geometry, convexity, and other courses with specialized content, I am sure recent ideas are treated. (I question, however, if these courses include as much applications-oriented material as they might.) However, the survey course, which is the course students are likely to see if they see any geometry at all, rarely includes significant recent material. As intellectually exciting as I find the axiomatics tale, I think we should not be teaching this any more. Rather, we should allow students to get this material on the side in much the way that now we expect them to learn the complement of axiomatics on the side. Recent pure and applied developments in all the far corners of our subject suggest the wisdom of doing this.

—Joe Malkevitch

I totally agree with Joe's statement that the axiomatic approach to Euclidean geometry is *not* the right way to go in a college geometry course, be it survey or otherwise. And I agree that more material on applications should be included. However, there is one tricky question, so far as the preparation of prospective high school teachers of mathematics is concerned. *If* they will still be expected to teach axiomatic plane geometry in high school, and *if* the only college geometry course that is taken by many of them does not cover this approach, then where are they to learn it?

—Vic Klee, University of Washington

The nature of geometry implies the nature of the geometry curriculum. From the material that has come in so far, it is clear that there is an enormous new breadth to geometry. I mean by that statement that there are parts to geometry that those of us who were educated twenty years ago never dreamed of. This is especially true of applications. Who would have dreamed that projective geometry would be applied in computer graphics, that differential geometry would have had so much success in control theory (or physics—but that was partially anticipated), or that there would be a subject called "computational geometry." I agree with Joe that "as intellectually exciting as I find the axiomatics tale, I think we should not be teaching ..." We need to construct a wider platform for geometry and geometric methods. The nature of geometry implies the nature of a geometry curriculum.

—Richard Millman, California State University

Joe Malkevitch wrote about the narrowness of calculus as an entry into modern mathematics. Also, as is generally true of teachers, "I am teaching you to be like me. This is the extent of my concept of a mathematician. So get a Ph.D. in mathematics, and specialize in hemi-demi-flipoids like I did."

Besides geometry, there are lots of subjects one could start out in and get a different view of the subject. Well-talked-over is discrete mathematics versus continuous mathematics. This has become a big issue because of computer science. Calculus as a first course in college-level mathematics is like Latin, Greek, and Hebrew in the classical education of yesteryear. You had to study these subjects and do well in them before you were deemed good enough to study anything else. Today, pre-meds have to take calculus before they are allowed to study pill-pushing. I think a course in discrete mathematics might be a better choice.

There is a very great need to formulate geometry courses for computer users. For example, computers are used in architecture to make pictures of buildings (and their shadows) which can be twisted and turned to different vantage points as they are exposed to sources of light with various intensities. You can see what a building looks like as you walk past it. Given a mathematical description of an object, how do you write the program which gives the view from one vantage point or another, and how does light from a given source change the picture? There is a lot of geometry in this.

Also, archeology courses (in Philadelphia) involve building sites brick by brick using the computer. There is a lot of geometry in this too. There are problems in oceanography where soundings have been made at certain points on the surface of a body of water, so the depth is known at this spot and at other spots too. Try to describe the surface of the ground under the water fitting these measurements. More geometry. Given a bunch of points in the plane, calculate the center and radius of the smallest circle that encloses all the points. How about the smallest pentagon? Calculate the convex hull of a given set of points. Classical problems in computational geometry. Image processing. Pictures of things. Get a picture of a regular icosahedron on the screen of a Macintosh. It might take somebody who knows lots of geometry quite a while to do this.

There is a need for a new course covering this type of geometry. The mother-of-all discrete mathematics books was written by Kemeny, Snell, and Thompson. After this great classic, a course was defined that spread to every college and university. The time is right for a book like this in geometry. I wonder what should be in the table of contents. Perspective drawing. Projections. Co-ordinate geometry. Hidden line algorithms. Other topics from computational geometry. Applications to various things—to medicine, architecture, archeology, making movies, art, you name it.

I think geometry fell from favor because of axiomatics. Because geometry is so close to intuition, we often feel we understand more easily without any formal model. One can solve geometric problems without thinking of the axioms. It is another game to prove things. Here the axioms play a role. To get a nice picture of the regular pentagonal dodecahedron on the computer screen which can be turned every which way with 'hidden lines' dotted as the solid turns is quite a project—but no axioms are needed.

—David Klarner, University of Nebraska

As I see it, we are trying to address the educational needs of at least three constituencies: the future teacher, the future mathematician, and the job seeker. At the same time there is discussion of "the course." Is it really possible to design a course that would be appropriate for all three groups? Or would it be more productive to design and press for the adoption of more geometry courses?

Another puzzle is that of the chicken and the egg. True, most teachers will be expected to teach axiomatics. But this may not be what they ought to be teaching—even in high school. I think the colleges and universities have to take the lead here. The Admissions Office here at Smith is frequently told that high schools follow the college's lead: if we require four years of mathematics, the schools will offer it; if we don't, the schools might bend to other pressures. The content of the geometry course is a more complicated issue, because the high schools feel they owe it to the students and their parents to prepare them for the SAT's, which seem to be rather hard to modify. But then if there is any hope of modifying them, it will have to be due to leadership from us. To summarize: the existing high school geometry curriculum should not deter us from teaching more appropriate geometry to prospective teachers.

—Marjorie Senechal, Smith College

I have lived in the Netherlands a total of three years and saw there a completely different educational setup. For one thing, there are about 13 million Dutch people, so the country is like a moderate sized state here. But the administration of the universities is a federal thing in the Netherlands. A percentage of GNP is used to fund them. Further, the universities are a federal resource, so they are used in administering the schools at all levels. So picking textbooks for the schools is done by people who know something about the subjects. Lots of top mathematicians get questioned about education at the lower levels. Since things are organized this way, people in the universities already understand their relationship to education, and they know what they are supposed to do about it.

Compared to the Netherlands, we suffer from a lack of definition of what we should do, and how to go about doing it. This lack of definition hurts when we try to settle what should be done about geometry. I still have my high school geometry book, and I must say the axiomatic presentation is awfully confused. Probably Peano's axioms for the natural numbers are easier to understand and use than the geometric axioms. In fact, Hilbert was the first to give a proper formulation of classical geometry. As far as reasoning in a rule-based system is concerned, there are lots of things you can do with chess problems. The disadvantage of axioms in geometry is that we are convinced that our intuitive notions are already adequate.

Well, with all this discussion, what do we the university experts think should be done? Is geometry the place to introduce axioms? As far as the history of ideas is concerned, that is where it started, and people ought to know about it. But there are simpler systems than Euclidean Geometry that might be used to illustrate the axiomatic approach. As far as problem solving is concerned, geometric or otherwise, I don't have much use for axioms.

Geometrize!

—David Klarner

The point that geometry is useful in new and unexpected ways has already been made well by Walter Meyer and David Klarner. We could describe most of this as "applications to computer science," but it really is applied geometry that happens to have been developed by computer scientists (for the most part) because they need it and it's not coming from mathematics. The question is, I think, how do we get on the bandwagon before it goes much further without us?

If you are interested in learning about some of what's going on in applied geometry, attend one of the upcoming SIAM meetings on "Computer-Aided Geometric Design" (CAGD). I went to one about four years ago. It was a real inspiration to see how vital a subject geometry had become, although you wouldn't know it if you looked at the mathematics curriculum at most universities. In addition to the CAGD people, there were talks on robotics, computational geometry, computational algebraic geometry, and automatic geometry theorem proving (Groebner basis theory), and polyhedral combinatorial optimization.

Two things about that meeting struck me particularly. The first was that the more interesting theoretical work being discussed had not been done in universities but in industrial research labs, General Electric and General Motors in particular, by people who seemed to have a strong background in various areas of geometry. Strangely enough, at least some few industrial research groups are able to take a longer-term view in the development of geometric methods. University groups that reported (for example in robotics or computer graphics) felt the need to have computer code or pretty pictures to show for their efforts.

The second aspect of the meeting that influenced my thinking came when Branko Grünbaum, at the end of his talk, pointed out to the assembled group that the pipeline for trained (and educated!) people to work in these exciting new areas was essentially empty. That nowhere in this country were there undergraduate students getting the necessary geometric background to be able to participate in these efforts. (Branko, please correct me if I've misrepresented what you said.) Since then I've thought about, talked about, but essentially done nothing about developing an upper-class course in applied geometry.

What I have in mind is something that already exists in many places in applied algebra. Several years ago I had a hand in introducing such an algebra course here. Since I taught it the first time, it has been given each year, now each semester, as well as during the summers (occasionally in multiple sections). It provides an introduction to the basic stuff of modern algebra (groups, rings, and fields) but does so in the context of more-or-less live applications: symmetry and Pólya enumeration, cryptography, error-correcting codes, etc. In addition to being popular with the students (some of whom are induced to take the "real" modern algebra course as a follow-up) it has been a real hit with the pure algebraists, who for a change like the idea of teaching something whose relevance is clear to the students. (It is so much of a hit that I'll only get to teach it myself for the second time next year.)

Applied geometry is a harder course to implement. There are no books (anyone interested?) and it's not clear what should be in them. As opposed to modern algebra, what is currently being taught in geometry in most places is irrelevant to these applications. The problem here is to assemble a list of geometric topics together with related live applications through which they can be introduced. Here the emphasis is on "live:" most students will not see the relevance of perspective drawing, Escher prints, and theoretical physics (as interesting as all of this is) to their ability to earn a paycheck. As crass as this may sound, it is exactly such considerations that leads most students to choose computer science over mathematics as a major.

The final issue I want to discuss was raised by Vic Klee and Marjorie Senechal, namely how to fit new ideas into the existing curriculum. I don't have any complete and coherent list of topics for a one-semester applied geometry course, but I think it's worth trying to make one. More likely, such a course will try to be at least a year long. So how to fit this into a mathematics major that now barely has any room even for axiomatic geometry (a logic course in disguise)? I don't propose we wait until the SAT's are reformed; I don't think we have the luxury of waiting that long.

Why don't we just require more mathematics of mathematics majors? It's my experience that mathematics has been one of the least demanding of majors in terms of requirements (compare with engineering!). This explains why in some places the mathematics major is popular with pre-meds, who need to take a lot of varied courses. When I was an undergraduate, I had enough free time to get the equivalent of a major in psychology on the side. Currently, I have an undergraduate advisee (who will be in Seattle next year as a mathematics graduate student) who managed to get a triple major (mathematics, operations research, and German). Surely having mathematics majors get a minor in computer science or operations research is a good thing, but why not offer the chance to minor in applied geometry (maybe some computer science and operations research students will join them). If you have ever done graduate admissions you know the difference in preparation of the average U.S. undergraduate mathematics major and that of mathematics students from virtually anywhere else.

—Louis Billera, Cornell University

Vic Klee raises a critical issue. If we change the survey course, whose historical audience has included a significant number of future high school teachers, how will such teacher trainees cope with what they are expected to teach currently in high school? This same point has troubled me when I have taught the survey course on and off for the last twenty years. My solution was to teach a lot of geometry that I felt was no longer the best choice for the students. Now, however, I think I see a solution to this problem, and it grows out of some of the teaching reform ideas, including writing across the curriculum. The idea is to have students in the survey course read on their own, not the details and the mathematics of the wonderful foundational story, but its history and an outline of the details.

There are a few books that deal with development of the interplay between the mathematics of the conception of space and the physicists' view. Two such are *Ideas of Space* by Jeremy Gray (Oxford University Press, 1980), and *Space Through the Ages* by Cornelius Lanczos (Academic Press, 1969). Both these books try to chart, in a conversational but not unmathematical way, the historical, mathematical, philosophical, and physical aspects of the interplay between geometry and space (in the physical sense). Another is H. Wolfe's book on *Non-Euclidean and Euclidean Geometry* (which is out of print, but which starts with a long introduction about these matters). I would ask the students to read one such book or collection of materials and write a "report" about it. This seems to kill two birds with one stone. It gives students an opportunity to see this material but more from a cultural perspective than a mathematical one, and it gives them an opportunity to read and write about mathematics in an historical framework.

I learned this type of thing in high school and in some college courses, whereas I learned the exciting ideas about tilings, Reuleaux triangles, curvature, and minimal surfaces, etc., etc., etc., from a math club. Why not reverse this? Put that other material at the fringes and get our eyes back on a kind of geometry that makes for a more meaningful current experience.

—Joe Malkevitch

I have taught a course on computational geometry four times now, twice to graduate students and undergraduates at Johns Hopkins, and twice to undergraduates only at Smith College. The course was offered in computer science departments, but some mathematics majors enrolled. Although what I cover may include more algorithms than most of you might prefer in an applied geometry course, I think it demonstrates the possibility of teaching such a course.

My experience is that the students can get *very* excited on these topics. The problems are motivated by applications, the relationships can be visualized, use of the computer makes it all concrete, and the easily-grasped open problems tantalize.

I cannot claim to be filling the empty pipeline that Lou Billera cites Branko Grünbaum as identifying, but I believe a trickle is being produced by those teaching computational geometry, mainly from within computer science departments. Several former students in these classes have maintained their interest in geometry for nearly a decade now.

If you will pardon the self-advertisement, I am currently writing a textbook for an undergraduate course in computational geometry. The prerequisites are basic programming, calculus, and a smidgen of linear algebra. Here are the five main topic headings, with a few sub-topics identified:

1. Partitioning Polygons. *Includes* art gallery theorems, polygon triangulation.
2. Convex Hulls. *Includes* 3-polytopes, Euler's relation.
3. Voronoi Diagrams. *Includes* Delaunay triangulations as projections of convex hulls, minimum spanning trees.
4. Arrangements of Lines and Planes. *Includes* connection to Voronoi diagrams.
5. Robotics. *Includes* shortest path algorithms, Minkowski sum, Collins' cell decomposition.

—Joe O'Rourke, Smith College

I have been reading people's comments with great interest. It does seem that part of the problem in thinking about geometry courses is that not only is there a diversity of constituencies but also there is quite a range of subject matter that suggests itself for undergraduate geometry.

In order not to get lost in all these possibilities, for the moment I would like to comment on a single case: students taking freshman and sophomore "paycheck" mathematics: calculus, linear algebra, and differential equations but not much beyond. These students are important, not only because they form the bulk of our mathematical customers in the undergraduate mathematics business, but also because they will go on to form the main body of scientifically literate citizenry.

While I would love to introduce a course called Geometry or Applied Geometry to be taken by all these students, I don't think we should ignore the possibility of putting back analytic geometry (or a modern substitute) into the standard introductory mathematics courses. This may not be a substitute for real geometry courses, but it can be a valuable adjunct and an improvement over the *status quo*.

I would like to report on one experiment that has gone on here at the University of Washington this year and then describe what I plan to do next year.

The third quarter of our freshman calculus course has been purged of other topics and is now devoted exclusively to multivariable calculus (differential calculus and multiple integrals but no div, curl, or Stokes theorem). The extra time is now devoted to a more extensive treatment of elementary vector geometry than was possible before; there is time to give some intermediate problems about lines and planes instead of just checking on the plugging-in of definitions. This course is taken by all the students on the main calculus track.

This year, as a companion to this third-quarter calculus course, two of my colleagues (Caspar Curjel and John Sylvester) have been running a one-credit computer lab course which concentrates on visualization in multivariable calculus. The students do not do what one normally thinks of as calculus problems. Instead they rotate pictures of 3-D objects (lines and planes and surfaces) with the mouse and try to analyze them, finding coordinates of points, directions of line segments, intersections, tangent planes, by looking rather than by using formulas. It is surprising how challenging this is.

For my own part, I have proposed to teach a similar lab next year to be attached to our linear algebra course, which suffers from a dearth of applications. My intention is to teach elementary linear algebra with a focus on geometric applications, looking at practical problems arising in engineering or computer science that can be solved using rotations, projections, transformations, etc. At the moment this is an experiment. In the future such a course in Linear Algebra with Geometric Applications could at least be an alternate to the regular linear algebra course if not a substitute for it.

To sum up, while I think we should be looking for ways to install geometry courses in the undergraduate curriculum, I don't think we should ignore the possibility of returning some of the geometric content to the standard calculus and linear algebra courses as well.

—Jim King, University of Washington

Although not everyone has been heard from yet, there seems so far to be little debate about the intellectual thrust of what a budding geometry movement should be: non-axiomatic, hands-on (including computers), with a strong flavor of applications and algorithms. The troublesome issues (so far) deal with strategies:

1. If we change geometry drastically at the college level, what happens to students who will have to teach the old stuff as high school teachers?
2. Should we put more geometry into the "paycheck" courses (great phrase!). Although the question was not raised in King's message, one might wonder whether that would be an alternative to a revitalized full course in geometry, or take the steam out of attempts to create one.

This leads me to the thought that we ought not to worry too much that we might upset some apple carts if we push forward in the wrong place. The curricular system is highly resistant to change. Anyone who has attended geometry sessions of mathematics meetings recently knows that there are many people teaching this subject who seem quite happy with very traditional ideas. We should push forward at both the high school and college levels; with separate courses or more geometry in old courses; with "applied geometry" and "computational geometry"—secure in the knowledge that the last thing that we are likely to do is to create a rapid revolution.

—Walter Meyer

The author introduces himself: First of all, I should confess that I am not a card-carrying geometer, but merely a defrocked algebraist who loves geometry. Moreover, I've been spending the last several years on a project which has been producing educational materials for school geometry (computer programs, computer-generated videotapes of three-dimensional figures, workbooks). Doris Schattschneider has been the real geometer on the project, but she lacks reliable bitnet to defend herself.

The conclusion thus far should be that you can safely ignore the mildly heretical views which I will soon begin expressing. Fair enough, but you should also educate me, since I frequently rush in where true geometers fear to tread. For example, the editors of the new MAA publication on *Visualization in Mathematics* conned me into doing an article on geometry, since they were having trouble finding the real McCoy. Moreover, the education project has been reasonably successful and Doris can't do everything, so I end up talking with teachers and administrators about school geometry. So argue especially hard when you disagree with me.

The author confesses to certain unhealthy tendencies: Brace yourselves, for I am going to say a kind word about axiomatics. The last few times I've taught geometry, I based the course on Greenberg's *Euclidean and Non-Euclidean Geometries.* I like very much examining some false proofs of the parallel postulate, and otherwise exhibiting the troubles with Euclid. It seems to me that one can well-prepare students to deal with the major repair job done by Hilbert and that it's very good training to get students to do some work with that axiom system. (It is much better training than in lots of algebra, for example, since geometry words carry such strong connotations.) Where better to show what axiomatics is really good for— a method for expressing mathematics well-suited to examination for gaps in reasoning and inconsistencies. Of course, one should feel honor-bound to make clear to students that this is how mathematics is frequently conveyed, but not how it's created.

On the other hand, only a sado-masochist would try to drag students through a rigid development from Hilbert's axioms, so I skip around when they start to turn purple. Greenberg is very good in not bringing in non-Euclidean models for a long time, and student minds are visibly stretched as they try to argue about hyperbolic properties from an axiomatic framework.

In sum, I think that if students are prepared to reason carefully within an axiom system, and if they see the necessity for doing so, it can be a very good thing. On the other hand, I readily admit that one can have too much of a good thing.

The biggest disadvantage to my (semi-)axiomatic approach is that it takes time, so I can't cover all the wonderful geometry I'd like to. (I've found this to be a problem with all the other approaches I've tried, too.) A final advantage of Greenberg's approach is that he very beautifully conveys the history of the discovery of non-Euclidean geometries, surely one of the most wonderful stories in all of mathematics. It's just grand to get in such splendid mathematics along with some history and philosophy. (Of course, I do take lots of side trips and detours to view some other geometry vistas.) I need to be convinced that I can do better for my general clientele than this approach, even though it is based on axiomatics.

—Eugene Klotz, Swarthmore College

I think our discussions so far fall into two categories: global issues and local issues. At the global level we have to consider how the nature of high school education has changed in the last thirty years (e.g., concern with larger percentage of successful graduates) and how college education has changed during the same period (e.g., larger volume of students, more concern on the part of students with how they will earn a living and how much they will earn, and the fact that computer science is a credible alternative for people with mathematical inclinations). Along with this there is the issue of what kinds of people study mathematics— their interests and motivations. I too have noticed the tendency for mathematicians to see the world in their own image. "Since I got turned on to mathematics by seeing Euclid in the tenth grade, it must be a good thing for everyone." Tradition is a very strong force at this level of analysis.

Local concerns are those dealing with how to restore geometry to a more central place in linear algebra and calculus; how to meet the needs of computer scientists and others with geometry courses taught within mathematics departments; how to treat the needs of future high school teachers; how to take advantage of the new technologies to teach geometry.

Several people have raised the idea of a new type of applied-oriented course, and tossed out possible topics, but were nervous about the great variety of possibilities. How about several independent attempts at a curriculum for such a course to see if there is a common core of topics people see as being significant?

Gene raises important issues in his comments on the fun, success, and value of doing axiomatics. Gene is too modest about what he has contributed to geometry. Ironically, as geometry has advanced and geometry courses have become traditional, there has been a growing gap between what geometers find interesting and what is being taught. Nevertheless, despite whatever value I used to see in teaching axiomatics for high school teachers or for mathematics majors with their eye on the Ph.D., I now believe that I can achieve my old goals and more with the kind of course I outline in "Geometry: Yesterday, Today, and Tomorrow."

—Joe Malkevitch

Having defended something which everyone else in the conference apparently dislikes, I'll now compound matters by attacking a consensus favorite. To salvage a wee bit of credibility here, let me relate some history. A long time ago I got the bug to teach applied algebra. My attempts were less than successful (I envy Louis Billera and his success). I had to cut back on the pure algebra which I normally covered, and in addition either I had to water down an example to the point where it appeared contrived to the students, or I had to load on so much detail from another subject that it only appealed to a small minority of the students. Since we already had developed enough traditional ways of turning off students, I abandoned this approach and went back to teaching "Groups, Rings, 'n Things," which was at least attractive to some very desirable students.

It seems clear to me that there should be real possibilities for courses with titles like Computer Science Geometry, but I question the viability of courses entitled Applied Geometry for Everyone. Beware the siren song of "relevance" which all too frequently attracts faculty but not students.

Last week our department had a discussion with students as to how we might improve our courses and make them more accessible. One faculty member suggested more applications. A student with impeccable taste voiced her concern that applications might be given in such a fashion as to convey the feeling that the mathematics wasn't interesting enough in itself. I'm on her side. I think that the contents of a mathematics course should be taken from the intersection of what students need to know, what is good for students, and wonderful mathematics. In geometry, I've always found the last subset so large that one can accommodate all sorts of changes to the first two.

From what I've seen, there's some very interesting new applied geometry. For example, I agree wholeheartedly with the favorable comments already made about computational geometry. The problem, as I see it, is that one has to be very careful—mediocre applied mathematics could force out some beautiful pure mathematics. This would be particularly unfortunate at a time when new possibilities are emerging for making a lot of classical geometry more accessible. But that will be the subject of another harangue.

—Eugene Klotz

I do not particularly wish to become embroiled in the pure mathematics—applied mathematics wars. From my point of view, such questions ultimately come down to matters of taste. What someone may refer to as "mediocre" applied mathematics may well seem aesthetically limited to me yet still be both non-trivial and important. After all, mathematics is used in the real world to the advantage of us all. If one decides to teach a geometry course and not include applications at all, I see no problem with this. I am a big believer in courses being most effective when the instructor is passionate about what he or she is teaching.

However, I see mathematics as having a flow and sense of direction as it grows. Some parts of what we as geometers know already is more or less important to the discovery and understanding of what we hope to know tomorrow. Sure there are new theorems about quadrilaterals to be discovered. But is this the best preparation for a student with an interest in geometry who is Ph.D.-bound? If the student is not Ph.D.-bound, but wishes to study actuarial science or work for an engineering firm, then there are criteria that could be applied to select optimal content for students at different levels.

—*Joe Malkevitch*

Last week I attended the conference in Orlando in honor of the 80th birthday of Howard Eves. There were over 150 participants from all over the country, ranging from middle school teachers to university professors and retired mathematicians. The talks represented Eves' fields of interest—geometry, pedagogy, history, and problems (as in problem columns in journals). It was a very pleasant and a very interesting meeting, with a lively, concerned audience, but I was struck by what seemed to me to be a lack of awareness of the scope of geometry and its burgeoning future.

I told them that mathematicians build gardens whose walls are axioms and definitions and propositions, and then busily cultivate the gardens, but often a plant escapes outside the walls and blooms more vigorously outside, or changes its characteristics, and we need to be aware of this. But nobody else was talking this way. On the other hand, it is clear that they all love geometry as they perceive it and are concerned about protecting its place in the curriculum. It seems to me that these folks—geometers all—need lots of accessible expository articles on contemporary geometry, and of course good texts. If this is correct, then it follows that we should be writing articles for journals like the *Monthly* and *Mathematics Magazine* and *Quant* and all the others, as well as writing interesting textbooks on a variety of geometrical topics. And the MAA should publish something like "Studies in Contemporary Geometry" aimed at the college teacher level (including those in community colleges). Maybe it already is doing something like this, but if it isn't, then it should.

Another useful thing—although a lot of work for somebody—would be a "Geometry Newsletter" with reports of conferences, lists of new books, discussions of models and visualization and other things, and a letters column.

It seems to me that we needn't worry much about axioms vs. applications, this vs. that. The world of geometry is enormous and we need *all* of it. The vs. problem only matters if you are trying to write *the* text for *the* course. But maybe in the future there will be many courses.

While working on *On the Shoulders of Giants* I was concerned about whether and how this material would reach the teachers. I don't know how it happened, but it did. Reports reaching me say they are reading it and talking about it. Maybe one outcome of this e-mail conference could be something like that—a collection of "strands" of geometrical thought. This would be a good theme for the MAA volume I mentioned a few paragraphs ago.

—*Marjorie Senechal*

The view from the mathematics education trenches is not all bleak when it comes to geometry. I would even expect that some of the favorable aspects might have a "trickle up" effect for college teaching. For example, there are some new and less-moribund texts starting to appear, most notably Mike Serra's *Discovering Geometry*. Moreover, some new computer programs are arriving which should have serious impact.

The first generation of these programs were those in Judah Schwartz's "Geometric Supposer" series. Although one can disagree with their procrustian Apple II design, they have nonetheless inspired a number of high school teachers and have an admirable user network.

It is the second-generation programs which I believe will be more revolutionary—programs such as the French "Cabri Geometrie" and our "Geometer's Sketchpad." (NB: Although I was involved in the production of the latter, the ideas involved are in the air—witness Cabri—and it is the type of program which should be of educational importance, not any particular one.)

What these programs can do is make Euclidean constructions easy, both for students and for teachers (some have unlimited undo and redo, so one can prepare a presentation and then easily go through it step-by-step). Beautiful "ancient" geometry which was always a pain to present at the elementary level (the 9-point circle, for example), can now stir student interest and not foment student revolt. These new programs not only facilitate the construction of figures that students find interesting, but they also allow the figures to be deformed so that students can search for the conditions under which a particular theorem holds. Some of these programs also show loci of points, envelopes of lines, and the like, which opens up lovely new vistas.

One effect of this will be the discovery of some new geometry—and the rediscovery of lots more—by mere high school kids. This is already happening with the Supposer, and the second-generation programs should create a deluge. Be prepared for endless questions as to the originality of elementary Euclidean results from high school teachers and students! (In particular, bet on a proliferation of loci problems, and Joe may even receive some new results about quadrilaterals—which he may not mind if they come from high school kids.)

One cloud I see on the horizon is that while the new programs encourage constructions and play, they don't seem to encourage proofs. For some students it seems to be enough to see that something holds; they still have to be convinced that it's important to ask "why?" I am busily looking for a construction which appears to hold on the computer screen, but which is in fact incorrect. (Thus far I am unsuccessful in geometry; for graphing programs, you can sometimes create a graph which looks just like $y = x$, but which is really a step function with many, many steps, but seen from a far distance. This can cause healthy mistrust in the infernal machines.)

The college geometry community would do well to see if they can provide direction as to what might constitute an appropriate revitalized high school geometry curriculum (if they can agree) since, if I am correct, more students may soon be attracted to geometry.

(They should also try some of the new software in classes. I've had a lot of fun with various hyperbolic models, inversion, projective generation of conics, transformations, etc.; so have my students. Both "The Geometer's Sketchpad" and "Cabri" are Macintosh programs.)

—Eugene Klotz

Gene Klotz asked for a construction which appears to hold on the computer screen, but which is in fact incorrect. Stan Wagon has a nice example of this in *Mathematica in Action*, p. 53. He plots the curve traced by the centroid of a Reuleaux triangle rotating inside a square, and it looks so like a circle that one might be convinced that it must be. But it is, in fact, composed of pieces of four congruent ellipses.

—Joe O'Rourke

Gene has raised the importance of computer environments for carrying out explorations of geometry. Not only are there nice tools such as the "Geometric Supposer" and the "Geometric Sketchpad," but there are also *Logo* and *Mathematica*. As Joe O'Rourke has pointed out, Stan Wagon has some lovely illustrations in his new book of how this can be done, and there is a large literature at the high school level, as Gene mentioned dealing with the "Supposer." I am a big fan of *Logo* and I feel it is unfortunate that more *Logo* is not done at the college level. There are other geometric toolkits in the areas of computational geometry and graph theory that are in the experimental stage of development. I feel that these tools can be employed to foster interest, exploration, and breadth.

—Joe Malkevitch

Focus Group Participants

tfb@cs.brown.edu	THOMAS F. BANCHOFF, *Brown University.*
billera@mssun7.msi.cornell.edu	LOUIS J. BILLERA, *Cornell University.*
frkc@flash.bellcore.com	FAN R.K. CHUNG, *Bell Communications Research.*
doyle@math.princeton.edu	PETER G. DOYLE, *Princeton University.*
rlg@research.att.com	RONALD L. GRAHAM, *AT&T Bell Laboratories.*
grunbaum@math.washington.edu	BRANKO GRÜNBAUM, *University of Washington.*
king@math.washington.edu	JAMES R. KING, *University of Washington.*
klarner@pythia.unl.edu	DAVID A. KLARNER, *University of Nebraska.*
klee@math.washington.edu	VICTOR KLEE, *University of Washington.*
klotz@cs.swarthmore.edu	EUGENE KLOTZ, *Swarthmore College.*
lee@ukma.bitnet	CARL W. LEE, *University of Kentucky.*
joeyc@cunyvm.cuny.edu	JOSEPH MALKEVITCH, Moderator, *York College, CUNY.*
adelphi@tigger.jvnc.net	WALTER J. MEYER, *Adelphi University.*
richard_millman@csusm.edu	RICHARD S. MILLMAN, *Calif. State Univ., San Marcos.*
jorourke@smith.bitnet	JOSEPH O'ROURKE, *Smith College.*
schattdo@moravian.edu	DORIS W. SCHATTSCHNEIDER, *Moravian College.*
senechal@smith.bitnet	MARJORIE SENECHAL, *Smith College.*

Reading List

1. Behnke, H.; Bachmann, F.; Fladt, K.; Kunles, H. (Eds.). *Fundamentals of Mathematics, Volume II: Geometry*. Cambridge, MA: MIT Press, 1974.

2. Berger, M.; Pansu, P.; Berry, J.; Saint-Raymond, X. *Problems in Geometry*. New York: Springer-Verlag, 1982.

3. Blackwell, W. *Geometry in Architecture*. New York: Wiley, 1984.

4. Davis, C.; Grünbaum, B.; Sherk, F. (Eds.). *The Geometric Vein*. New York: Springer-Verlag, 1981.

5. Eves, H. *A Survey of Geometry, Revised Edition*. Boston, MA: Allyn and Bacon, 1972.

6. Kappraff, J. *Connections: The Geometric Bridge Between Science and Art*. New York: McGraw-Hill, 1991.

7. Lindquist, M., and Shulte, A. (Eds.). *Learning and Teaching Geometry, K–12*. Reston, VA: National Council of Teachers of Mathematics, 1987.

8. Lord, E., and Wilson, C. *The Mathematical Description of Shape and Form*. Chichester: Ellis Horwood, 1986.

9. Meschkowski, H. *Unsolved and Unsolvable Problems in Geometry*. New York: Fredrich Ungar, 1966.

10. Morris, R. (Ed.). *Teaching Geometry: Studies in Mathematics Education, Volume 5*. Paris: UNESCO, 1986.

11. Stehney, A.; Milnor, T.; D'Atai, J.; Banchoff, T. *Selected Papers on Geometry*. Washington, DC: Mathematical Association of America, 1979.

Appendix A: COMAP Geometry Conference

In recent years, there has been a tremendous surge in research in geometry. This surge has been the consequence of the development of new methods, the refinement of old ones, and the stimulation of new ideas both from within mathematics and from other disciplines, including computer science. Yet during this period of growth, education in geometry has remained stagnant. Not only are few of the new ideas in geometry being taught, but also fewer students are studying geometry.

In March 1990, a group of college and university researchers and educators in geometry met to assess the directions of education and to make suggestions for invigorating it. These individuals represented a wide variety of branches of geometry as well as a wide spectrum of institutions. Discussions ensued on the causes of the decline in geometry education and on the steps that might be taken at all grade levels (K–graduate school) to energize the teaching of it. Special attention was given to the content of the survey course in geometry taught in many universities and colleges. This course has historically been taken by a large number of prospective high school teachers, and thus setting new directions for this course offers the hope of exposing future mathematics practitioners to new ideas in geometry, as well as for laying the basis for future changes in lower grades.

Despite the varied points of view expressed by the individuals who attended the conference, there was a broad core of common views, which, if implemented, can have a significant effect on geometry. This common core of views and recommendations is presented below.

These recommendations and the following article "Geometry: Yesterday, Today, and Tomorrow" by Joe Malkevitch are reproduced with permission from *Geometry's Future*, the proceedings of a March 1990 conference sponsored by COMAP, Inc. (57 Bedford Street, Suite 210, Lexington, MA 02173).

Conference Recommendations

Future directions for the teaching of geometry (especially for implementation in the college/university survey course):

- Geometric objects and concepts should be studied more from an experimental and inductive point of view rather than from an axiomatic point of view. (Results suggested by inductive approaches should be proved.)
- Combinatorial, topological, analytical, and computational aspects of geometry should be given equal footing with metric ideas.
- The broad applicability of geometry should be demonstrated: applications to business (linear programming and graph theory), to biology (knots and dynamical systems), to robotics (computational geometry and convexity), etc.
- A wide variety of computer environments should be explored (*Mathematica*, *Logo*, etc.) both as exploratory tools and for concept development.
- Recent developments in geometry should be included. (Geometry did not die with either Euclid or Bolyai and Lobachevsky.)
- The cross-fertilization of geometry with other parts of mathematics should be developed.
- The rich history of geometry and its practitioners should be shown. (Many of the greatest mathematicians of all time: Archimedes, Newton, Euler, Gauss, Poincaré, Hilbert, von Neumann, etc., have made significant contributions to geometry.)
- Both the depth and breadth of geometry should be treated. (Example: Knot theory, a part of geometry rarely discussed in either high school or survey geometry courses, connects with ideas in analysis, topology, algebra, etc., and is finding applications in biology and physics.)
- More use of diagrams and physical models as aids to conceptual development in geometry should be explored.
- Group learning methods, writing assignments, and projects should become an integral part of the format in which geometry is taught.
- More emphasis should be placed on central conceptual aspects of geometry, such as geometric transformations and their effects on point sets, distance concepts, surface concepts, etc.
- Mathematics departments should encourage prospective teachers to be exposed to both the depth and breadth of geometry.

Appendix B: Geometry: Yesterday, Today, and Tomorrow

by Joe Malkevitch, YORK COLLEGE, CUNY

Despite the increased pace of exciting developments in both the theory and applications of geometry in the last 40 years, it appears that less geometry is being taught in college today than was taught in the recent or distant past. The purpose of this paper is to examine this "paradox" and to study how the teaching of geometry in colleges affects what geometry is and can be taught in high school, grade school, and graduate school mathematics.

Geometry in Mathematics Departments Today

A perusal of recent college catalogues show mathematics departments listing (though not always regularly offering) a variety of geometric-based courses: Graph Theory, Differential Geometry, Convex Sets and Geometric Inequalities, Combinatorial Geometry, Projective Geometry, Topology, etc. In addition to courses such as these, many mathematics departments offer a survey course in geometry under a variety of titles. These include College Geometry, Euclidean and Non-Euclidean Geometry, Topics in Geometry, Modern Geometry, Geometric Structures, etc. It will be convenient to refer to the first type of course as a Geometry Course and the second type as a Survey Course. (Geometry also enters the curriculum in a variety of other courses including Calculus, Linear Algebra, Combinatorics, etc.) Although it is rare to require either of these types of courses of students majoring in mathematics, it is not uncommon for many mathematics departments to require a Survey Course or some Geometry Course from those mathematics majors planning to teach mathematics in secondary schools. This type of requirement reflects the fact that the "traditional high school curriculum" includes a year of study of geometry in the tenth grade. Thus, the geometry taught in college is closely tied through teacher preparation to the geometry taught in pre-college mathematics. To explain the decline in the teaching of geometry in college requires a digression.

Why Students Major in Mathematics in College

Most college mathematics majors fall into one of the following groups: students planning to enter graduate school to start in a program of doctoral studies in mathematics, students planning careers as high school (or sometimes intermediate school) teachers, students planning to pursue careers relating to computing, students planning actuarial careers, students planning to enter an "applied" master degree program (this is usually a terminal degree that does not result in the student pursuing the doctorate degree), and "others." Especially at colleges in large metropolitan areas, high school mathematics teachers have traditionally constituted a significant portion of the total number of mathematics majors. With the downturn in mathematics majors that was seen in many colleges during the period 1972-1988, one saw a dramatic reduction in the number of students preparing for careers as secondary school teachers. This reduction is ostensibly attributable to several phenomena. First, the dramatic decrease in the number of students in the school system during the period meant that many teachers currently in the profession were laid off. Second, the dramatic over-supply of mathematicians, engineers, etc., in the post-Apollo period made students wary of majoring in these subjects, and the high salaries paid in the computer science siphoned away many students with interests in mathematics. Third, the salaries of high school mathematics teachers relative to other professions that potential teachers of mathematics could enter became eroded.

When the downturn in mathematics enrollments in general and mathematics secondary school teachers in particular hit our colleges, the effect on the teaching of geometry courses was especially extreme. This is clearly related to the fact that the major group of students taking survey courses were future high school teachers. Even for geometry courses, loss of enrollment in high school teacher audiences resulted in decreased offerings. It is unfortunate that this diminished exposure to geometry for mathematics majors has come at a time of tremendous dynamism for geometry itself.

What is Geometry?

Before continuing with more detailed discussions, it may be useful to explain how the term geometry has been and will be used in this essay. In attempting to clarify what is meant by the term geometry, it is clear that the word "geometry" means different things to different audiences, including subgroups of the mathematics community itself.

To lay people, geometry is the study of the space and the shapes that they see in the world around them. Most lay people's exposure to geometry is the simple material on classification of shape that they learn about in grade school and the exposure to "pseudo-axiomatic" geometry in high school. Much of high school geometry is still highly concerned with the axiomatics and the proving of Euclidean theorems in a manner that has come to be described as two-column proofs. This refers to a series of statements and the reason for the statements in a second column. In recent years, there has been a growing movement toward a more "inductive" approach to geometry, spurred on in part by the development of such software packages as the "Geometric Supposer." However, this movement has been nearly exclusively concerned with the metric properties of triangles, quadrilaterals, and circles. Thus, to the non-mathematician, geometry has a very narrow meaning. Obviously, "geometry" has much richer connotations to members of the mathematics community.

However, even within the mathematics community, geometry means a surprisingly diverse number of things to different people. To some, geometry refers to those portions of mathematics (and mathematical physics) that deal with the mathematical structure of space, thereby involving a large variety of deep mathematical tools such as operator theory, partial differential equations, and Lie groups. To others, it refers to differential geometry and the topology of manifolds. Yet other groups think of it as meaning (though not exclusively) the emerging body of ideas dealing with discrete geometrical structures. As diverse as the meaning of the word geometry is, a remarkably large portion of the subject can be introduced and profitably pursued with a minimum amount of background and formal study of mathematics. In this sense, geometry differs greatly from other parts of modern mathematics such as functional analysis, ring theory, logic, algebraic topology, etc.

Here the word geometry will be used in its very broadest sense of all aspects of mathematics where visual information, diagrams, models, and understanding of space are involved or put to use. For an attempt to catalogue the breadth of ground entailed by this viewpoint, see Malkevitch [8]. It is noteworthy that a variety of rapidly emerging areas within mathematics and computer science have a major geometric component. In order to see how geometry fits in the college curriculum of the future, it will be useful to examine the traditional relationship between geometry and other parts of mathematics.

Geometry's Relation to Mathematics

It is interesting to note that although many areas of mathematics have first been developed in geometric form, these areas have often matured when they were algebratized. Examples include synthetic Euclidean geometry, projective geometry, block designs, catastrophe theory, etc. As important as geometry is both to geometers and mathematics, as a separate discipline, it has never been in the mainstream of mathematics, once mathematics as a subject for study was institutionalized in universities and colleges. In a pre-World War II university or college, during the period when the roots of the current renaissance in geometry were being laid out at the research level, there were fewer Geometry and Survey

Courses being taught than would have been the case from 1960-1975. Thus, a university during the 1920s or 1930s would have had courses in Analytic Geometry, Solid Analytic Geometry, Projective Geometry (perhaps in both synthetic and algebraic versions), and (old style) Differential Geometry. The wealth of geometry courses listed (though often untaught) at the college and university of today were uncommon then. In fact, at that time, no explicit survey course in geometry existed. (No equivalent of Howard Eve's pioneering *Survey of Geometry* (1963) with its curious forward- and backward-looking collection of topics existed before the War. The niche for high school teachers, trained then in "normal" schools or colleges, was filled by courses such as College Geometry or Modern Geometry. For a sample of the books of that era see Eves [1, p. 115]. Courses on convex sets, graph theory, groups and geometry, etc., virtually did not exist.)

Today, a standard introduction to mathematics for a graduate student pursuing a doctorate degree consists of a year of Real and Complex Analysis, a year of Abstract Algebra, and a year of Topology (with geometric aspects of the subject not necessarily emphasized). The teaching of topology often serves the role of hand-maiden for parts of Real and Complex Analysis. Judged by the dissertation titles that one sees listed in recent years by the American Mathematical Society, geometry is a relatively minor field at the fingers of most research. (Perhaps symptomatic of geometry's problems is that in the new 1990 mathematics subject classification list, the rapidly emerging area of computational geometry receives no listing.) The qualifier/preliminary examination system in place at most (especially as implemented at large) graduate schools discourages entry into "fringe" areas such as geometry. Thus, in a certain very real sense, the study of geometry has not been in the mainstream of the training of professional mathematicians: those majoring in mathematics in college and going on to pursue doctoral studies in graduate school.

Since there are not enough individuals who call themselves geometers to go around, most survey courses in geometry are taught by individuals with a narrow base of geometrical knowledge. Such individuals rely heavily on the geometry texts in print in teaching the Survey Course since teaching a course based on readings and on their own knowledge base imposes a heavy preparation burden. (Geometry Courses are taught by the one member of the department who got the course listed in the catalogue in the first place, are taught by a "draftee," or fall into disuse.

As noted before, many parts of mathematics have been developed in geometric form. Furthermore, a true renaissance of geometry has occurred in recent years. Examples of this ferment in geometric ideas include: the development of a new branch of mathematics, computational geometry; exciting breakthroughs in understanding the geometric structure of space (with resulting heavy cross fertilization with workers and ideas in mathematical physics); breakthroughs in the study of the mathematics involved in tiling problems for both the plane and higher dimensional spaces; an explosion of geometric ideas related to the theory of graphs with application to many areas of mathematics and operations research; dramatic new developments in the theory and application of the theory of knots; exciting connections between developments in the theory of dynamical systems and the geometry of sets (fractals); dramatic uses of geometrical methods in image recognition and processing; and use of geometric methods in the control and motion planning for robots and robot arms, to mention but a few of the most visible examples. This listing could easily be extended. Hence, it is increasingly unfortunate that both teachers (already teaching and new ones being trained) and future researchers have not had available to them a vehicle for being exposed to the exciting new developments in geometry. Though geometric thinking itself may not be taught as part of the mathematical mainstream, geometry and geometric thinking is "infiltrating" mainstream mathematics more than ever before.

Geometry and Teacher Training

If American citizens are not to be raised as geometric illiterates, teachers in our grade schools and high schools will have to be broadly trained geometrically themselves. We have already examined the trend that new high school mathematics teachers entering our schools are few in number and have had less opportunity to be exposed to geometry than high school teachers of earlier generations.

Many experiments are now being conducted to try to develop specialists to teach mathematics K–6. The need for "mathematics specialists" has been raised by the resistance of traditionally trained K–6 teachers to new developments and teaching methods in grade school. (Traditionally trained teachers in elementary school usually take a single course in mathematics as part of their teacher preparation. This course concentrates almost exclusively on the development of thinking about the base 10 number system, associated problems in addition, subtraction, multiplication, and division, and on measurement. This course rarely mentions any ideas in the area of geometry beyond simple taxonomy of simple shapes.) Emerging programs that urge specialists for elementary school to major in subject areas in college, as more reasonable preparation for teaching in grade school, will wind up subjecting such students to the very narrow type of geometry course now taught as a Survey Course in our colleges. One of the few positive trends to note is that many teachers, both those planning to teach in high school or pre-high school environments, are being forced or encouraged to study the computer language called *Logo*. Creative use of the *Logo* language can permit students to be exposed to a wide range of open-ended, exploratory experiences with geometry.

Clearly, the Survey Course in geometry will play a large role in the exposure of future teachers to geometry. This is likely to become more so if future grade school mathematics specialists take this type of course. Thus it seems both wise and necessary for the mathematics community to significantly revamp the Survey Course. Such a change will be a service not only for future teachers and their students but for future researchers as well.

Goals in Changing the Survey Course as it Currently Exists

In attempting to change the content of the Survey Course, there are a variety of reasonable goals. Among these is the possibility of significantly changing the content of what is taught in high school by giving future high school teachers preparation in the geometry that might be part of a future high school geometry curriculum. Another goal is to encourage larger groups of students with interests in areas related to mathematics (e.g., computer science and engineering) to explore the many advantages that would accrue to them in being more broadly versed in geometric ideas. (The self-contained and quick starting nature of geometry makes this feasible.) A final goal might be to provide a rich variety of geometric concepts and tools for future research mathematicians both in traditional as well as emerging areas of mathematics, and to encourage more future research mathematicians to work in the area of geometry by exposing students to easily accessible unsolved problems.

Benefits of a Newly Constituted Survey Course

Although clearly geometry deserves to be studied for its own sake, many important objectives of mathematics study in general can show from studying geometry. Below is a partial list of some of the benefits of a revised geometry Survey Course (listed in random order):

- To show how geometric mathematics is affecting modern life (i.e., compact disk recorders, CAT scans, HDTV [high definition TV], image processing, richer understanding of the geometry of space, robots, new types of maps, etc.).
- To encourage visual thinking and reasoning (use of diagrams and models as modes of thought and problem solving).
- To learn the interplay of pure and applicable ideas (e.g., error-correcting codes and sharp pictures of Uranus and Jupiter, know theory to study DNA, etc.).
- To learn the distinction between the mathematics of geometry and the geometry of physical space.
- To show the rich history of geometry as a subject and the connection between geometry and other disciplines outside of mathematics such as philosophy and physics.
- To show how computers and specific software environments can be an aid to geometric thinking.
- To foster better writing, verbal, and communication skills when dealing with technical ideas.
- To illustrate how ideas in mathematical modelling are of value in a geometrical setting, and how geometric thinking is a tool for the mathematical model builder (i.e., use of graph theory to study problems in making deliveries to discrete locations, say oil to homeowners).
- To learn how ideas developed for one application of mathematics are often transportable to other situations (e.g., getting a fire truck to a fire quickly and designing efficient paths for robots in a workspace).
- To obtain experiences in problem posing and problem solving.
- To illustrate domains in which experiments can be done in mathematics and have students carry out such experiments (e.g., soap bubbles, tilings, mirrors to study symmetry).
- To expose students to a variety of unsolved problems in geometry.
- To learn how one part of mathematics makes contributions to other parts (e.g., the interplay between algebra and geometry, and combinatorics and geometry).
- To illustrate how basic concepts such as distance, function, volume, etc., are of use in a geometric setting.
- To illustrate the power of abstraction, special cases, and the use of symbolism.
- To learn what a mathematical proof means and to give examples of such proofs. (Note: there is no reason, however, to restrict the domain of such proofs to theorems that appear in Euclid or similar results.)

Content for a New Survey Course

In attempting to design a new Survey Course in Geometry a variety of principles could be applied. Among these are that basic geometrical concepts and methodologies should be represented, that modern applications should be shown, that breadth as well as depth be respected, that a variety of geometric proof techniques be shown, and that a variety of different types of geometrical objects be examined. In addition to teaching a course based on significantly new content, I believe that the mathematics community should take advantage of new computer technologies (computer environments such as *Logo* or *Mathematica*) and the use of videotape. For example, many applications of geometry are best introduced to a student in visual form using videotape rather than in written form. Appendix II shows various ideas for development of a video applications library to support existing and future text materials used in the teaching of geometry.

As a brief perusal of Malkevitch [8] quickly reveals, an exhaustive look at geometry in a semester sequence is not realistic. There is just too much attractive and important material. Any specific geometer is likely to have a somewhat different collection of topics and ordering for teaching these topics for a survey course from another geometer. However, I believe there is widespread agreement that the current course must be changed, moved in a direction away from axiomatics, and that any new course have a "core" of principles and content. In Appendix I, I have listed one of many possible approaches to both the content and organization of a new survey course that I have considered. Implementation of such a course will, I believe, be a major step toward attaining greater geometric literacy for teachers, the lay public, and mathematicians as well.

References

1. Eves, H. *A Survey of Geometry.* Boston, MA: Allyn and Bacon, 1963.
2. Grünbaum, B. "Shouldn't We Teach Geometry?" *Two-Year College Mathematics Journal*, 12 (1981) 232–238.
3. Grünbaum, B. and Shephard, G. *Tilings and Patterns.* New York: W.H. Freeman, 1986.
4. Hadwiger, H.; Debrunner, H.; and Klee, V. *Combinatorial Geometry in the Plane.* New York: Holt, Rinehart and Winston, 1964.
5. Hargatti, I. (Ed.). *Symmetry.* New York: Pergamon Press, 1986.
6. Henderson, K. (Ed.). "Geometry in the Mathematics Curriculum." *Thirty-sixth Yearbook.* Reston, VA: National Council of Teachers of Mathematics, 1973.
7. Lindquist, M. (Ed.). "Learning and Teaching Geometry." *1987 Yearbook.* Reston, VA: National Council of Teachers of Mathematics, 1987.
8. Malkevitch, J. "Geometry in Utopia." *Geometry's Future.* Arlington, MA: Consortium for Mathematics and Its Applications, 1990.
9. Richards, J. *Mathematical Visions.* San Diego, CA: Academic Press, 1988.
10. Stehney, A.; Milnor, T.; D'Atai, J.; and Banchoff, T. *Selected Papers in Geometry.* Washington, DC: Mathematical Association of America, 1979.
11. Walter, M. (Ed.). *Readings in Mathematical Education: Geometry.* Derby, England: Association of Mathematics Teachers.

Appendix I: Geometry Tomorrow

Outline of some of the major topics to be covered in a geometry course of the future (listed in random order):

- Combinatorial ideas vs. metric ideas
- Convexity
- Geometry of physical space (relation to axiomatics)
- Graph theory ideas
- Computational geometry ideas
- Symmetry, polyhedra, and tilings
- Visual thinking
- Area and volume (Bolyai-Gerwin, Hadwiger, Banach-Tarski)
- Dynamical systems and fractals
- Differential ideas
- Applications

- Role of dimension
- Proof tools: induction, infinite descent, examples, constructions, divide and conquer, etc.
- Isomorphism concepts
- Geometric transformations
- Digital geometry
- Packing and covering problems
- Lattice point problems

Assumed prerequisites: year of calculus or principles of mathematics course, and knowledge of matrix notation and multiplication (but not necessarily of linear algebra).

Unit I

A. What is Geometry?
  - i. Geometric pearls
    - a. Bolyai-Gerwin theorem
    - b. Euler's traversability theorem
    - c. Art gallery theorem (Fisk's proof)
    - d. Helly type theorems
    - e. Curves of constant breadth
    - f. Distance realization problems
    - g. Euler's polyhedral formula
    - h. Penrose tiles
    - i. Pick's theorem (lattice points)
    - j. Desargues' theorem
  - ii. Visual thinking
    - a. Value of drawing diagrams
    - b. Value of constructing models
    - c. Geometry experiments (soap bubbles, etc.)
    - d. Computer environments
B. Different Approaches to Geometry
  - i. Difference between metric geometry and combinatorial geometry
  - ii. Axiomatic geometry and the geometry of physical space
  - iii. Isomorphism concepts
  - iv. Historical role of parallelism
    - a. Two space
    - b. Three space
    - c. Four space
    - d. Space-time
    - e. Surfaces embedded in three space
    - f. Dimension
  - v. Deductive vs. inductive approaches to the study of geometry
  - vi. Geometry and the computer
  - vii. The relation of geometry to algebra and other parts of mathematics
C. Proof Tools of the Geometer
  - i. Induction
    - a. On number of objects
    - b. On dimension

    - ii. Infinite descent
    - iii. Constructions
    - iv. Algebra
    - v. Arguments based on symmetry

Unit II

A. Types of Geometric Structures (Graphs, Planes, Spaces, Block Designs, Convex Sets)
B. Graph Theory
  - a. Traversability
  - b. Trees
  - c. Coloring problems
  - d. Planarity
  - e. Matchings
  - f. Network algorithms (shortest paths, flows, minimum-cost spanning trees, etc.)
C. Planes
  - a. Affine planes
  - b. Projective planes
  - c. Hyperbolic planes (infinite and finite examples)
  - d. Role of Desargues' "statement"
D. Space
  - a. Euclidean, projective, hyperbolic space
  - b. Axiomatics and geometry of space
E. Block Designs
F. Convex Sets
  - a. Helly, Radon, and Cartheodory's theorems
  - b. Minkowski addition
  - c. Curves of constant breadth
  - d. Geometric inequalities (isoperimetry)
  - e. Lattice point problems
  - f. Packing and covering problems

Unit III

A. Geometrical Transformations (viewed not as an approach to the theorems of Euclidean geometry, but for their own sake)
B. Transformations and Their Relationship to Space
C. Transformations and Their Relationship to Metric Properties (i.e., congruence)
D. Geometric Transformations Viewed Geometrically
E. Geometric Transformations Viewed Algebraically

Unit IV

A. Symmetry and Regularity Polygons
  - a. Plane polygons
  - b. Convex polygons
  - c. Self-intersecting polygons
  - d. Packing and covering problems

B. Tilings
    a. Tilings with regular polygons
    b. Tilings with convex polygons
    c. Symmetry properties of tilings
    d. Aperiodic tilings
    e. Penrose tilings

C. Polyhedra
    a. Regular polyhedra
    b. Archimedean polyhedra
    c. Combinatorial properties of polyhedra
       i. Euler's formula
       ii. Steinitz's theorem
    d. Minkowski addition aspects of polyhedra
    e. Graphs of polyhedra
    f. Tilings in space

D. Symmetry Groups
    a. Symmetry groups of tilings, patterns, fabrics, etc.

Unit V

A. Area and Volume
B. Equidecomposability
C. Role of Archimedes' Axiom
D. Squaring the Circle
E. Banach-Tarski Paradox
F. Dynamical Systems and Fractals

Unit VI

A. Computational Geometry
B. Triangulations
C. Voronoi Diagram
D. Sweep Line Methods
E. Convex Hull
F. Principles of Design for Geometric Algorithms

Unit VII: Topological Ideas

A. Geometry of Surfaces
    a. Orientability (Moebius Band)
    b. Torus
    c. Klein Bottle

B. Knots
    a. Geometric transformations of knots
    b. Classification of knots

Unit VIII: Geometric Optimization Problems

A. Linear Programming
B. Isoperimetry
C. Packing and Coverings
D. Network Optimization

Unit IX: History of Geometry

A. Geometry in the Ancient World
B. Geometry During the Renaissance
C. Geometry Up to the 20th Century

D. Geometry in the 20th Century

Note: There should be biographical material about the great contributors to geometry, including, where possible, portraits or photographs.

Unit X: Applications of Geometry

Applications should probably be sprinkled in and included in an integral manner with the other parts of the materials being developed. However, here are some particularly topical areas that might be mentioned (see Appendix II for additional examples):

A. Robotics
B. Computer Vision
C. Computer Graphics
D. Solid Modelling
E. Operations Research

Note 1: Unsolved problems in geometry would be mentioned throughout the course.

Note 2: For bibliographic references in support of a wide variety of classical and recent topics, see Malkevitch [8].

Appendix II: Ideas for a Videotape

Edge Traversal

Situations:
    Curb inspecting
    Street sweeping
    Garbage collection
    Mail delivery
    Advertising circular delivery
    Painting line down center of roads
    Snow removal
    Parking meter collection and enforcement
    Police or museum guard patrol routes
    Pipe, wiring, or duct inspection

Mathematics:
    Graphs as models
    Euler's traversability theorem
    Chinese Postman Problem
    Johnson and Edmond's algorithm
    Deadheading and repeated edges

Practitioners:
    U.S. Postal Service
    Sanitation Department
    Department of Parking Enforcement
    University Operations Research Departments (MIT, Maryland, Stony Brook)
    AT&T Bell Laboratories; Bell Communication Research

Vertex Traversal

Situations:
    Meals on wheels
    Deliveries to supermarkets, restaurants, etc.

Garbage pickup from industrial sites
Machine inserter schedules
Computer solution of jigsaw puzzles
School bus routes
Camp pickup routes
Parcel post delivery and pickup
Pizza delivery
Special delivery of mail
Pickup of coins from pay telephone booths

Mathematics:
Graphs as models
Hamiltonian circuits in graphs
Traveling salesman problem
Asymmetry of costs
Complexity
K-opt methods
Greedy algorithms
Vehicle routing problems
Clarke-Wright algorithm

Practitioners:
Sanitation Department
U.S. Postal Service
Federal Express
Parcel Post
School Boards
Camps
University Operations Research Departments (MIT, Stony Brook, Maryland)
AT&T Bell Laboratories; Bell Communications Research

Voronoi Diagrams

Situations:
District planning
Drainage regions
Market structure (anthropology)
Robot motion planning

Mathematics:
Computational geometry
Perpendicular bisector
Convex set
Convex hull
Concurrence, concyclic points
Line sweep algorithms

Practitioners:
University Mathematics and Computer Science Departments (Smith College, Courant Institute, University of Illinois, Princeton, Rutgers)

Robots (Motion Planning)

Situations:
Industries which employ mobile robots
Planetary surface exploration

Mathematics:
Graphs as models
Visibility graphs
Shortest path algorithms
Minkowski addition
Parallel domains
Vision

Practitioners:
General Motors, Ford, Chrysler, etc.
Universities (MIT, Yale, Courant Institute (NYU), Stanford)
AT&T Bell Laboratories

Note: Other aspects of robotics also involve geometrical ideas. These include the local motion planning of the gripper of a stationary robot.

Bin Packing

Situations:
Machine scheduling (independent tasks)
Organizing computer files on disks
Advertising breaks
Want advertisements in newspapers

Mathematics:
Packing problems
Heuristic algorithms
Measures of efficiency
Time space tradeoffs
Complexity
Simulation

Practitioners:
Operations Research Departments (Berkeley)
AT&T Bell Laboratories

Distances

Situations:
Car travel
Urban distance
Biology (evolutionary trees)

Mathematics:
Taxicab metric
Abstract properties of distance
Sequence comparison
Levenshtein distance

Practitioners:
AT&T Bell Laboratories

Shortest and Longest Paths

Situations:
Fire truck and ambulance routing
Building construction
Space program (flight planning)
Robot motion planning

Mathematics:

Graphs, digraphs, and weighted graphs and digraphs
Dijkstra's algorithm
Critical path method

Practitioners:
Operations researchers

Minimum-Cost Spanning Trees

Situations:
Synthesis of communication networks
Road planning

Mathematics:
Graphs as models
Trees
Spanning trees
Kruskal's algorithm
Prim's algorithm
Greedy algorithm

Error Correcting Codes

Situations:
Compact disk players
Computer codes
Space programs
HDTV

Mathematics:
Binary sequences
Distance (Hamming distance)
Matrices
Information content

Practitioners:
Compact disk manufacturers (Philips)
Universities (California Institute of Technology, MIT)
AT&T Bell Laboratories

Coloring Problems

Situations:
Scheduling committees, final examinations, railroads
Fish tanks and animal confinement patterns
Maps
Placement of guard in art galleries

Mathematics:
Graphs as models
Vertex colorings
Face colorings
Edge coloring
Complexity

Practitioners:
Universities with graph theory specialists
AT&T Bell Laboratories; Bell Communications Research

Data Compression

Situations:
Image transmission and storage
Text transmission and storage

Mathematics:
Binary numbers
Digitalization of text and images
Huffman codes
Fractal methods

Practitioners:
Universities (MIT, Georgia Institute of Technology)
AT&T Bell Laboratories; Bell Communications Research
NASA

Geometric Transformations and Symmetry

Situations:
Computer graphics
Analysis of fabrics
Analysis of archeological facts
Cartography
Analysis of art (Escher paintings)

Mathematics:
Group theory
Functions and transformations
Strip groups
Wallpaper groups
Color symmetry

Practitioners:
University mathematicians

Unfolding Polyhedral Surfaces

Situations:
Catching a spider on the wall of a cube
Drawing a map of a spherical surface
Unfolding the surface of the brain
Layouts for packages

Mathematics:
Development of polytopes
Projection mappings
Distance on polyhedral surfaces

Block Designs

Situations:
Drug testing
Agricultural productivity
Scheduling workers
Scheduling tournaments

Mathematics:
Finite geometries
BIBD's
Orthogonal Latin Squares

Practitioners:
  Universities (Ohio State University, CAL Tech)
  AT&T Bell Laboratories

Mathematical Programming

Situations:
  Blending gasolines
  Blending juices
  Manufacture of processed foods
  Scheduling
  Shipment of goods
  Vehicle routing
  Hospital management
  Portfolio management

Mathematics:
  Linear programming
  Integer programming
  Linear inequalities
  Solution of liner equations
  Network flows
  Transportation problem

Practitioners:
  Universities (Rutgers, Princeton, Stony Brook)
  AT&T Bell Laboratories
  Oil companies, airlines, car companies, defense industries

Art Gallery Theorems

Situations:
  Surveillance in museums, banks, and military installations

Mathematics:
  Convex sets
  Types of polygons
  Triangulations
  Colorings

Practitioners:
  Computational geometers

Euclidean Geometry

Situations:
  Length of carpet remnants
  Time remaining on a partially used tape

Mathematics:
  Geometry of the circle (circumference)
  Areas and perimeter concepts
  Isoperimetry

Alternate Organization

The situations above are organized by mathematical theme. Other approaches also exist, in particular, showing applications of geometry to a particular subject area. Several examples of this are given below:

Applications of Geometry to Business:
1. Traversability problems
2. Minimum-cost spanning trees
3. Facility location problems
4. Coloring problems (scheduling problems)

Applications of Geometry to Medicine:
1. CAT scanners and other medical imaging systems
2. Kidney stone machines
3. Brain mapping studies

Applications of Geometry to Biology:
1. Structure of the gene (intersection graphs, interval graphs)
2. Food chains, niche spaces, competition (intersection graphs)
3. Ecology (fractals)
4. Shape of biological forms (isoperimetry)

Applications of Geometry to Chemistry:
1. Quasicrystals (Penrose tiles, crystallography)
2. Dynamics of chemical reactions (dynamical systems)

Applications of Geometry in Communications:
1. Synthesis of communication networks
2. Vulnerability of communication networks
3. Phone exchange systems
4. Error-correction methods and codes
5. Data compression (compression of text and images)
6. Digitalization of images
7. Image processing (filtering, etc.)

Applications of Geometry in Social Science:
1. Analysis of fabrics, designs, and pottery (anthropology), groups, symmetry patterns
2. Kinship systems (anthropology), graph theory
3. Mobility (sociology), Markov chain digraphs
4. Equilibrium analysis (economics), dynamical systems

Environmental Mathematics

Environmental Mathematics

Ben Fusaro

SALISBURY STATE UNIVERSITY

Introduction

Environmental mathematics is an emerging field, waiting to be delineated. Whatever its final definition, a synthesis of a subject whose central issue is the survival of life on this planet and a subject that is crucial for modern science and technology will command attention.

> The environment is the most important problem facing the human race today, and the only effective response involves educating the general population. I can see no better place to begin the educational process than in the mathematics curriculum.
>
> —*Dean Hoover, Alfred University*

> This project is perhaps the most valuable one that mathematical educators can be involved in at this time.
>
> —*Marty Walter, University of Colorado, Boulder*

Richard Schwartz of College of Staten Island, a pioneer whose efforts to raise environmental consciousness via introductory mathematics goes back fifteen years, thinks it is essential that we marshal our efforts to "...make mathematicians and others much more aware of the critical nature of environmental problems."

The term "environmental mathematics" made its first appearance in a national publication in the April 1990 issue of *Focus* ("Solving Environmental Problems: Where Are the Mathematicians?", Ben A. Fusaro and Marcia P. Sward). The topic was first part of a national AMS–MAA program as a panel discussion at the San Francisco meeting in January 1991. The success of this panel led the 1992 Program Committee to call for a proposal for increased environmental activities at the Baltimore meeting. Subsequently, environmental mathematics was called to the attention of the full AMS–MAA community by being featured in announcements for this meeting. One nice side effect of this publicity is that mathematicians who were not initially part of our Focus Group wrote in with suggestions. Thus, there will be names in this document that do not appear among the list of Focus Group participants at the end. (These volunteers turned out to be very important. Our e-mail system, faltering almost from the beginning of the project, got worse as time went on.)

Setting the Stage

I tried to set the stage by providing an implicit definition of environmental mathematics (EM), by listing some characteristics, and by suggesting how EM should be introduced to the curriculum.

Environmental mathematics needs to be distinguished from courses in ecosystems, eco-modelling, mathematical ecology, and similar subjects. Ecology is usually defined as the study of the relation between organisms and their environment. The "eco" courses tend to be in biology departments, or are taught as upper-division or graduate applied mathematics courses. Our intention is that environmental mathematics refer to undergraduate activity (including high school).

It could be said that environmental mathematics has the same relation to the environment as engineering mathematics has to engineering. This definition by analogy gives the topic ample scope, and received the implicit approval of the Focus Group participants, with one word of caution:

> If the word "mathematics" is emphasized in the title "environmental mathematics," then the environmental science students will not take it. If the environmental part is emphasized, then it doesn't sound like a mathematics course."
>
> —*Lothar Dohse, University of North Carolina, Asheville*

A crucial difference between "eco-" courses and environmental mathematics is that the former can be—and often are—taught in a clinical "white coat" fashion. An article that was part of a nationally-distributed announcement illustrates this point. The author labored for several paragraphs to convince the reader how important it was for scientists to classify and study all endangered species before they disappear. Only at the end of the article was there a "by the way" that perhaps something ought to be done to keep these species from disappearing.

Tom Hallam of the University of Tennessee, Knoxville sounded a warning. He was concerned that *too* strong of a partisan approach could lead one to imposing one's views on students. Bob McKelvey of the University of Montana had sounded a similar warning at the EM panel discussion in San Francisco. He suggested that one had to be careful not to let a strong environmental concern intrude on—and perhaps distort—a scientific message. McKelvey emphasizes that working in environmental mathematics often leads one to deal with policy, and therefore controversy.

Initially, the main effort in environmental mathematics should be directed at developing courses suitable for general education. Here is where the biggest impact could be made. The effort could then be shifted to introducing environmental topics into sophomore or junior modelling courses. The topics might deal with pollution, recycling, (so-called) timber management, water quality, energy, etc.

Characteristics of Environmental Mathematics

The concern about partisan teaching suggests a need to clarify the "white-coat science" issue. No one wants "PC"-type pressures as have surfaced in several prominent universities. However, environmental mathematics needs to include an awareness of the environmental cost of implementing a model, and beyond that, a general concern for environmental degradation. The subject cannot avoid dealing with *values*.

Indeed, competing value systems are often built into the problem.

> Resolving the relative harm to those with differing views is often the *point* of the problems we address: economics vs. wildlife, jobs vs. spotted owls, erosion and logging vs. salmon runs.
>
> —*Rollie Lamberson, Humboldt State University*

Our populace is becoming less able to evaluate or assess the trade-offs between alternatives. It would be exceptionally useful if the environmental courses would be able to open the eyes of the non-mathematics major to the evaluation of alternatives.

—*Lee Seitelman, Pratt & Whitney, Connecticut*

The health of the environment is so critical for us all, wildlife and *homo sapiens* alike, that we dare not hold the subject at arms length—we need to become engaged. A few years ago there was a mathematical conference on ecological modelling at which tuna was served for lunch. A sampling of the opinion of a few participants indicated that they were either unaware of the relation of yellow-fin tuna consumption to a protected species or, even worse, did not see what our lunch menu had to do with the topic of the conference. Are these the kind of role models we want to teach students the mathematics of the environment ...?

The challenge is to have mathematics serve as an instrument for modelling and for raising environmental consciousness in "a setting where scientific explorations can be pursued openly."

General Education and Entry-Level Courses

There was fairly strong agreement that first efforts should go to developing introductory courses or materials for general education. The next level of effort should be to develop materials for lower-level modelling courses, for calculus courses, and for other beginning courses for majors. Since part of the concept is to use mathematics to develop environmental awareness, the earlier the subject is introduced, the better.

Two participants, Fusaro and Schwartz, have developed materials for general education courses and have been teaching them for nine and sixteen years, respectively. Fusaro has been using an environmental text [2] for background and notes. Supporting packaged software is available [3]. Schwartz has published a text, *Mathematics and Global Survival* [5].

Marty Walter has recently proposed a course called "Mathematics for the Environment" with only high school algebra as a prerequisite. S.S. Dalal suggested a course built entirely around energy, noting that "energy is of great interest to students."

Students are to explore different forms of energy and write papers which are presented in class for general discussion. Students are given about fifteen topics to choose from. Some examples are: solar energy, wind energy, renewable energy, energy from biofuels, small-scale hydropower systems, and municipal resource recovery.

—*S.S. Dalal, Embry-Riddle Aeronautical University, Florida*

Others are working in different ways to introduce environmental issues into first year courses. Christopher Schaufele and Nancy Zumoff of Kennesaw State College, Georgia presented an MAA paper "Applications of Algebra to the Environment" at the Orono, Maine summer meeting. They offered a replacement for college algebra called "Earth Algebra" [4].

Another interesting development was revealed in a letter from Geoffrey Beresford of Long Island University, New York. His letter started with "I am writing a calculus textbook" [Gasp!], but continues with "and am very interested in finding undergraduate-level applications to environmental science. I am making a special effort in my book to raise environmental consciousness." [Ah ...!]

Modelling Courses

Most of the Focus Group participants described modelling courses in which environmental materials were introduced to varying degrees. Lothar Dohse has taught a Mathematics Modelling course in which most of the student projects were environmental or biological. Rollie Lamberson reports that Humboldt State began in 1980 to offer a sophomore-junior course with emphasis on biological and environmental problems.

Robert Wenger of the University of Wisconsin, Green Bay comes from a department that has been actively working on environmental problems for about fifteen years. He makes two important points. First, since environmental problems are interdisciplinary, mathematicians need to make the effort to interact with other disciplines. Second, sophisticated models or techniques are usually not required. He also notes that little is available in textbook form and suggests a series of UMAP-type modules focused on environmental problems. (There are a few modules in the COMAP 1992 catalog: Nos. 207, 607, 610, 628, 653, 670, 675, and 688.) Generally, modelling courses dealt with open-ended projects, required student projects and presentations, and encouraged team efforts.

Conclusion

One of the striking aspects of most courses on environmental mathematics is the similarity of classroom management techniques and the alignment of instructional styles with contemporary recommendations about effective teaching. Respondents in our study made use of almost all of the following: experiential learning, classroom presentations, cooperation or team efforts, interdisciplinary approaches, open-ended problems, and term projects. It seems that in environmental mathematics courses, form follows content.

Events indicate that mathematics and environmental science are headed for a marriage. It is a marriage we should encourage as professionals and reflect in the curriculum as teachers. Getting environmental mathematics into the curriculum will be a parallel process, but the initial emphasis should be on introductory courses. An eventual goal might be the development of masters degree programs in environmental mathematics, perhaps along the lines of the Environmental Systems program at Humboldt State University. At all levels, these experiences need to be accompanied by an environmental consciousness in the context of open inquiry.

Annotated Bibliography

1. Harte, John. *Take a Spherical Cow: A Course in Environmental Problem Solving*. Mill Valley, CA: University Science Books, 1988.
   In this paperback the author covers a wide variety of environmental problems, from sulfur in coal to pollution of lakes. There is a great emphasis on the modelling process. The author relies heavily on knowledge of the natural sciences; many of the problems make use of calculus or differential equations. Neither calculators nor computers play much of a role in the text.
2. Odum, H.T. and Odum, E.C. *Energy Basis for Man and Nature*. New York: McGraw-Hill, 1978.
   This is a fascinating paperback introduction to Odum's energy analysis and language. Some knowledge of biology is required to appreciate the applications. Seven basic models are studied. BASIC code is provided in an Appendix.

3. Odum, H.T. and Odum, E.C. *Computer Minimodels and Simulation Exercises*. Gainesville, FL: Center for Wetlands, 1989.

   This paperback is intended to teach students to diagram, simulate, and experiment. There are 45 diagrammatic models, as well as BASIC code for the Apple II, IBM PC, and Macintosh. It supplies some background for each model, but does require the user to be familiar with the subject matter. The first nine models are general, the next seventeen are biologically or ecologically oriented, and the last nineteen are applications to economics.

4. Schaufele, C. and Zumoff, N. *Earth Algebra*. Glenview, IL: Harper Collins, 1991.

   The authors were led to this work by trying to answer the questions "What makes College Algebra boring to practically every student on this planet?", and "What can be done to make College Algebra more interesting?" They deal with these questions by extensive use of environmental contexts and by designing a text that exploits the power of a graphing calculator. These authors are not "white coat" scientists. They view environmental issues as vital, and their concern is reflected throughout the text.

5. Schwartz, R.H. *Mathematics and Global Survival, Second Edition*. Needham Heights, MA: Ginn Press, 1990.

   This is a freshmen-level paperback with a wide variety of environmental applications. It uses such issues as hunger, pollution, and resource allocation to motivate basic calculations and descriptive statistics. It is the basis for a course, Mathematics and the Environment, that the author has taught since 1975.

Focus Group Participants

coles@math.utah.edu	WILLIAM J. COLES, *University of Utah.*
mcozzens@note.nsf.gov	MARGARET B. COZZENS, *Northeastern University.*
dohse@unca.bitnet	LOTHAR A. DOHSE, *Univ. of North Carolina, Asheville.*
e3fafusaro@sae.towson.edu	BEN FUSARO, Moderator, *Salisbury State University.*
mhood@cosmos.acs.calpoly.edu	MYRON HOOD, *Calif. Polytechnic Univ., San Luis Obispo.*
fhoover@ceramics.bitnet	DEAN W. HOOVER, *Alfred University.*
rollie@calstate.bitnet	ROLAND LAMBERSON, *Humboldt State University.*
lewiss@cwu.bitnet	SCOTT M. LEWIS, *Central Washington University.*
mckelvey@stolaf.edu	STEPHEN MCKELVEY, *St. Olaf College.*
rhssi@cunyvm.cuny.edu	RICHARD H. SCHWARTZ, *College of Staten Island.*
	LEON H. SEITELMAN, *Pratt & Whitney, Connecticut.*
msward@athena.umd.edu	MARCIA P. SWARD, *Mathematical Association of America.*
walter@euclid.colorado.edu	MARTIN E. WALTER, *University of Colorado, Boulder.*
	ROBERT B. WENGER, *Univ. of Wisconsin, Green Bay.*

Appendix A: Three Course Outlines

Mathematics and the Environment (Marty Walter)

Each student selects a problem for a semester-long project, with monthly progress reports required. Exams take the form of oral presentations to the class.

Warm-up Exercises:
- How many cobblers are there in the U.S.?
- How far will a drop of water spread on water?

- How large was the Ancient Asteroid that (perhaps) killed the dinosaurs?
- Population growth.
- What fraction of the total annual plant growth was eaten by humans last year?
- How much sulphur was put in the air by burning coal last year?

Modelling Tools:

- Steady-state box models and resident times.
- Thermodynamics and energy transfer.
- Chemical reactions and equilibria.
- Non-steady-state box models.

Open-Ended Problems:

- Acid rain.
- Mobilization of trace materials.
- Carbon cycle tracing.
- Global warming and the greenhouse effect.
- Optimal harvesting.
- Steady-state population in China.
- Road-killed rabbits in Nevada.

Earth Algebra (Christopher Schaufele & Nancy Zumoff)

Both of us are "environmentalists" and began development of the course "Earth Algebra" in hopes of achieving two goals: to educate students at an early college level on environmental issues, and to demonstrate the effectiveness of mathematics as a decision making tool. Early evaluations indicate success in each category.

"Earth Algebra" is aimed at beginning college students. It utilizes all concepts from a traditional college algebra course to simplistically model environmental data, to make more decisions regarding predicted events, to evaluate alternative energy sources, and to formulate recommendations for changes which will improve environmental conditions. The course is very focused; everything in the text is relevant to the issue of global warming. This puts college algebra in a context which most students are already aware of, and to a certain degree, interested in; it gives relevance to mathematics, and hence generates interest in and purpose to its study.

This project is supported by grants from NSF and FIPSE; Harper Collins will publish the text in late 1992.

BRIEF TABLE OF CONTENTS

I. Carbon dioxide concentration and global warming.
    1. Introduction: The greenhouse effect and global warming.
    2. Atmospheric carbon dioxide concentration.
    3. On the beach, ... or, what beach?

II. Factors contributing to carbon dioxide build-up.
    4. Carbon dioxide emission from automobiles.
    5. Carbon dioxide emission from energy consumption.

    6. Carbon dioxide emission from deforestation.

    7. Total carbon dioxide emission functions.

III. Accumulation of carbon dioxide.

    8. Increases in atmospheric carbon dioxide concentration.

    9. Factors contributing to atmospheric carbon dioxide concentration.

IV. Social factors.

    10. People.

    11. Money.

V. Save the planet!

    12. Changing energy demand.

    13. Cost and efficiency of alternative energy sources.

    14. What can you do to save the planet?

Environmental Mathematics (Ben Fusaro)

The emphasis is on computational, qualitative, and visual mathematics. All modelling is done by a seven-step process, moving from the visual and qualitative to the computational (calculators and BASIC). The students solve differential equations but they are called "Flow Equations." There is a major project that is done (preferably) by teams of two students. Course outline:

- Systems and Diagrammatics
- Energy and Entropy
- Energy and Growth
- Simulation of Models
- Energy Flow and Money Flow
- Production and Diversity

Appendix B: Mathematicians Develop New Tools to Tackle Environmental Problems

by David L. Wheeler, THE CHRONICLE OF HIGHER EDUCATION

Idling cars spewing fumes, northern spotted owls seeking nesting sites in diminishing plots of old-growth forest, and molecules of sulfur dioxide settling through the branches of the human lung: Such events would not strike most scientists as inherently mathematical. But mathematicians using graphs, equations, and their own brand of abstract thinking have been involved in each of those problems and are seeking a larger role in other environmental research.

"Environmental mathematics is an attempt to get mathematicians to connect again with the natural world," says Ben A. Fusaro, a professor at Salisbury State University and the chairman of the Mathematical Association of America's new committee on mathematics and

Reprinted with permission from *The Chronicle of Higher Education*, 38:20 (January 22, 1992) pp. A7, A10–A11.

the environment. Mr. Fusaro was an organizer of a series of talks, workshops, and discussions of environmental issues at the Association's joint meeting with the American Mathematical Society here this month.

Models of the Natural World

Mr. Fusaro says mathematicians can help environmental researchers by building models of the natural world and by seeking out both the variables and the things that do not change, or "invariants," in a living system. Most important, he says, mathematicians can help environmental researchers by finding the internal structures that link many different phenomena. One differential equation, for example, describes both the bouncing movements of a weight that is suspended from a mattress spring and the oscillations of an electrical current in a radio.

Robert McKelvey, a professor of mathematical sciences at the University of Montana who created a mathematical model for the northern spotted owl population in the Pacific Northwest, says that mathematics is needed to help set specific environmental policies because old methods of arriving at such decisions have failed. Arriving at a policy decision by placing a dollar value on both the costs and the benefits of an action, for instance, can't work if a dollar value can't be assigned to one or even both sides of cost-benefit calculations, Mr. McKelvey says.

Assessing the Costs

The timber industry is eager to point out the costs in jobs and dollars imposed by a logging ban in the mature forests where the northern spotted owl lives. But environmentalists claim no price can be set on the loss of the reclusive owl, which is protected by federal endangered-species legislation, or of the forests where it lives, which took hundreds of years to form.

Both sides, says Mr. McKelvey, are trying to preserve something that cannot be assessed in dollars: One wants to preserve a way of life tied to logging and the other a forest undisturbed by humans.

Mathematicians, working with economists, psychologists, and others, have developed a formal theory of making decisions with multiple conflicting objectives, known as multiple-criterion decision theory. That method, and others developed by mathematicians, could provide clear outlines of environmental problems, Mr. McKelvey says. "In the end you can't find a magic formula that tells you what to do," he says, "but the trade-offs can be made more explicit."

Mathematicians are accustomed to trying to comprehend uncertainty, Mr. McKelvey says, while many policy makers are afraid of it. He points to the controversy over global warming: Policy makers, he says, "are so frozen by their conservative natures that if they don't know what is going to happen, they don't do anything."

Good and Bad Years

In his own work, Mr. McKelvey has estimated how the portion of old-growth forests that is saved from logging in the Northwest will affect the chances of losing all the northern spotted owls. The owls prefer to nest under the canopy created by the tall trees in the old-growth forest, apparently because they have a better chance of escaping attack from predators there.

With a computer model of the owl population, Mr. McKelvey simulated a series of good and bad years for owls. In good years, owls have plenty of food—chiefly small rodents—and search for new nesting sites and breed. In bad years, the population stays stable or declines. The model randomly creates good and bad years and simulates 250-year periods.

Mr. McKelvey's model used information gathered by biologists, such as the amount of territory a pair of nesting owls requires. After thousands of computer runs simulating various combinations of good and bad years, the model showed that a critical threshold exists for the survival of the owl: When less than 20 percent of the old-growth forest is saved the chances of the owl's survival drops sharply.

Although many may argue about the model's assumptions or the precise location of the threshold, the knowledge of the threshold's existence is a valuable contribution, Mr. McKelvey says. Likewise, he says, the mathematical models can help biologists determine what data are needed to improve such predictions.

Pollutants in the Lungs

Sometimes mathematics is used to model aspects of the biological world that scientists would have difficulty studying in any other way.

At the Center for Mathematics and Computation in the Life Sciences and Medicine at Duke University, mathematicians are trying to determine what happens to pollutants that enter the human lung. The configuration of the lungs in other species is so different from humans' that laboratory animals cannot be used to study the health effects of pollution in humans, says Michael C. Reed, a professor of mathematics and director of the Duke center.

Because experimental surgery on humans is out of the question, Mr. Reed says mathematical models are one of the few tools available to help scientists understand what doses of pollution different parts of the lung will receive when breathing different concentrations of pollutants.

To solve the problem, mathematicians must first understand lung physiology. The sacs at the end of the lung, Mr. Reed says, have an enormous surface area: 80 to 100 square meters, the largest area in the body that is exposed to the outside air. "This is an enormous surface just sitting there and waiting to be injured," Mr. Reed says.

The branches of the lung—tubular bronchioles—are protected by mucus that, in conjunction with the cells lining the lungs, sweeps many pollutants up and out of the lung. The mucus coat thins near the junctions of the lung's branches and is missing completely at the junctions themselves.

Duke researchers have created two-dimensional models that can simulate portions of the human lung, the thickness of the mucus lining, and the motion of the air and the pollutants that it carries into the lung during breathing. The models have helped the scientists discover that the edges of the sacs, near the bronchioles, are likely to receive high concentrations of asbestos fibers when a person is breathing air containing them.

"The equations we are applying have been known for a hundred years, but the techniques of solving those equations are changing all the time," says Satish Anjilvel, a mathematician and an assistant professor of medicine who is working on the lung models. Understanding the deposition of pollutants in the lungs will keep many applied mathematicians busy for at least a decade, says Mr. Reed.

In the lower regions of the lungs, Mr. Anjilvel says, the flow of air is considered to be "laminar" and can be described exactly by standard equations. But in the upper regions of the lungs and in the nose, the flow is turbulent and cannot be simulated exactly by existing equations, he says.

Ending Traffic Jams

At Rutgers University, Fred S. Roberts, a professor of mathematics, conducts research designed to reduce the pollution from automobiles by eliminating traffic snarls. Mr. Roberts uses mathematical tools known as interval graphs to time traffic lights to prevent unnecessary idling of automobile engines.

The interval graphs, originally developed in 1959 to deduce the shape of genes, represent overlapping lines in a figure as points on a graph. Each point stands for an overlap: If two parallel segments don't overlap, there is no point on the graph for them.

In applying the graphs to traffic problems, mathematicians represent traffic flow that can occur simultaneously as points on the interval graphs. Traffic motion that cannot occur simultaneously, such as cars turning left and cars coming from the opposite direction, would not appear as points on the graph.

Mathematicians can search for the largest possible "clique," or cluster of points on an interval graph, to find how to move traffic efficiently.

The mathematical problem then expands to determine how to order the phases of green lights and how long each phase should be. As adjacent lights and surrounding streets are added, the problem becomes an increasingly challenging one for mathematicians.

Another problem on which Mr. Roberts has worked is the design of one-way street patterns. Many cities have adopted one-way streets to move traffic more quickly. But the patterns, which also use graph theory, must be designed without making it too difficult to drive from one place to the other. Transportation officials might, for instance, ask mathematicians to arrange the pattern of one-way streets to make the longest trip that anyone has to take as short as possible.

Mathematicians do not have a way of computing the solution to that problem for all patterns. "Unfortunately, we come very quickly to the forefront of mathematical knowledge," says Mr. Roberts.

Reaching for Quantitative Literacy

Reaching for Quantitative Literacy

Linda R. Sons

Northern Illinois University

The Focus Group on Quantitative Literacy addressed three key issues—why, what, and how. A lengthy introduction written by Donald Bushaw provides a response to the first question. The report of the Focus Group itself is devoted to a discussion of "what" and "how," based on experiences from many different institutions. Appendices provide information on research, commentary on course materials, and extracts from the Focus Group e-mail conversation.

Introduction: Why Quantitative Literacy?

by Donald Bushaw, Washington State University

There seems to be wide agreement that a well-educated citizen should have some significant proficiency in mathematical thinking and in the most useful elementary techniques that go with it. In western civilization, the idea goes back at least to classical times, when four (the "quadrivium") of the seven liberal arts considered essential for the education of a free citizen were essentially mathematical. The role of mathematics was enlarged by the Enlightenment, by the Industrial Revolution, and by many events in modern science, technology, business, and the rapid intellectual evolution of humanity generally.

In recent years, amidst intense scrutiny and sometimes harsh criticism of the whole educational system in the United States, one group after another has expressed itself on the point. A representative statement (here considerably abbreviated) appears in the influential report *Everybody Counts: A Report to the Nation on the Future of Mathematics Education* (National Academy Press, 1989, pp. 7–8):

> To function in today's society, mathematical literacy—what the British call "numeracy"—is as essential as verbal literacy...Numeracy requires more than just familiarity with numbers. To cope confidently with the demands of today's society, one must be able to grasp the implications of many mathematical concepts—for example, change, logic, and graphs—that permeate daily news and routine decisions...Functional literacy in all of its manifestations—mathematical, scientific, and cultural—provides a common fabric of communication indispensable for modern civilized society. Mathematical literacy is especially crucial because mathematics is the language of science and technology...

An emphasis on the expanding importance of general education in mathematics beyond high school was made over twenty years earlier, in the COSRIMS report *The Mathematical Sciences: A Report* (1968), p. 56:

> The impact of science and technology has become so significant in our daily life that the well-educated citizen requires a background in the liberal sciences as well as the liberal arts. It has long been recognized that mathematical literacy is an important goal of all liberal education. But in current education this training often stops at the secondary-school level. With the increasing quantification of many of the newer sciences, the impact of high-speed computers, and the general expansion of the language of mathematics, it becomes increasingly important for the college graduate to have some post-secondary training in mathematics...

Or consider the following words from *The Mathematics Report Card: Are We Measuring Up?* (Educational Testing Service, 1988, p. 9):

> Looking toward the year 2000, the fastest-growing occupations require employees to have much higher math, language, and reasoning capabilities than do current occupations. Too many students leave high school without the mathematical understanding that will allow them to participate fully as workers and citizens in contemporary society.

Those who have been pleading for more nearly universal quantitative or mathematical literacy have not all been mathematicians, by any means. Consider the words from *50 Hours: A Core Curriculum for College Students* (National Endowment for the Humanities, 1989, p. 35):

> To participate rationally in a world where discussions about everything from finance to the environment, from personal health to politics, are increasingly informed by mathematics, one must understand mathematical methods and concepts, their assumptions and implications.

These statements and many others like them add up to an interesting challenge, and since about half of American colleges and universities have no general mathematics requirement for graduation, the challenge is clearly not being met.

There have been encouraging signs of improvement in recent years, but optimism can be premature. As these words are being written, it was just announced by The College Board that the average quantitative score on the SAT has taken another downward turn, after more than a decade without any decrease.

We have been speaking of *mathematical* attainments. The term "quantitative literacy" has so far appeared only in the title. Whether there is a real difference between "quantitative literacy" and "some significant proficiency in mathematical thinking and in the most useful elementary techniques that go with it" is a matter of debate. Sometimes the term "quantitative literacy" is a virtual euphemism for some level, usually ill-defined, of accomplishment in mathematics. (How unfortunate that some people should consider it expedient to use a euphemism for "mathematics"!) At other times "quantitative literacy" is used much more broadly, to include logic, linguistics, and other subjects that have at least a relatively formal character, even if they are seldom or ever taught in mathematics departments.

Here we shall adopt the point of view that "quantitative literacy" primarily concerns mathematics, broadly understood. It is not an entirely fortunate term. For one thing, much of modern mathematics, even at elementary levels, is not distinctively quantitative; for another, "literacy" suggests both facility with *letters* and a possibly very low level of accomplishment. The term "numeracy" is shorter, at least. Most, if not all, of what will be said here will apply whichever reasonable interpretation of the term "quantitative literacy" is adopted.

It may be useful to enumerate some of the principal reasons for expecting quantitative literacy of educated people. The list that follows is surely not complete, and the items in it are not independent; but it directs attention to some of the major areas in the broad range of "Why study mathematics?"

- Mathematical thinking and skills are of great value in *everyday life*. "Other things being equal, a person who has studied mathematics should be able to live more intelligently than one who has not. And, up to a point at least, the more mathematics studied, the more intelligent the life should be" (NCTM, *A Sourcebook of Applications of School Mathematics*, 1980, Preface).

- One of the classic reasons for studying mathematics is that it strengthens *general reasoning powers*, for instance by developing problem-solving skills. While the research literature is ambiguous on this point, many thoughtful people are convinced that it is true in some sense.

- Quantitative literacy at varying levels is clearly needed in *preparation for further study* in many academic and professional fields. It is reliably estimated that the majority of undergraduates would be required to take a course or courses in the mathematical sciences for this purpose even in the absence of a general graduation requirement of this kind.

- Increasing amounts of mathematics are needed in an increasing number of *careers*. ... "More and more jobs—especially those involving the use of computers—require the capability to employ sophisticated quantitative skills. Although a working knowledge of arithmetic may have sufficed for jobs of the past, it is clearly not enough for today, for the next decade, or for the next century" (*Moving Beyond Myths: Revitalizing Undergraduate Mathematics* (National Academy Press, 1991, p. 11). And students, even college seniors, often do not know what careers they will enter, or where their career paths will lead them. A quantitative literacy requirement helps to hold some doors open.

- Many adults, and especially college graduates, are very likely to assume positions in their communities and in professional organizations where quantitative literacy (e.g., the ability to deal intelligently with statistics) will come into play and may even be essential for effectiveness. A quantitative literacy requirement can thus be expected to enhance the quality of *citizens*.

- Anyone who does not have a mature appreciation of mathematics misses out on *one of the finest and most important accomplishments of the human race*. A quantitative literacy requirement, sensibly defined, will contribute to the spread of that appreciation.

- Society can ill afford to under-develop *latent mathematical talent*. For many students the activities leading to satisfaction of a quantitative literacy requirement can be revelatory, inspiring them to consider for themselves careers in mathematics or mathematics-related fields.

- The fear of mathematics that is often called *math anxiety* or *mathophobia*, besides stunting the cognitive development of those who suffer from it, tends to communicate itself from one generation to the next, in the home, and elsewhere. It is usually learned, not in-born, and a quantitative literacy course or courses, if competently and compassionately taught, can be powerfully therapeutic against it. (Certain learning disabilities do seriously impede the learning of mathematics, but the number of people affected by these disabilities is small. Reasonable accommodations, for legal as well as humanitarian reasons, should be made for such students.)

Even if, as many thoughtful people believe, the educational process that finally produces college graduates should be regarded as seamless, practical considerations require that some line should be drawn between the pre-college part and the college part, or in other words between the secondary part and the tertiary part. The present report is sponsored by the Mathematical Association of America, which by its charter is concerned with "collegiate mathematics," so is concerned mainly with the college part.

The term "remedial" (or "developmental"), as applied to a college mathematics course, has a definite meaning only where there is a clear understanding of where pre-college mathematics leaves off and collegiate mathematics begins. There are various opinions about where this line may be. However "remedial" is defined, the volume of remedial instruction to college students has certainly increased in the past several decades. According to *A Challenge of Numbers: People in the Mathematical Sciences* by Bernard L. Madison and Therese A. Hart (National Academy Press, 1990, p. 29):

> In fall 1970, college enrollments in remedial courses constituted 33% of the mathematical sciences enrollments in two-year colleges and by 1985 had increased to 47%. In four-year colleges and universities, remedial enrollments constituted 9% of the mathematical sciences enrollments in 1970 and had increased to 15% by 1985.

In spite of the volume of resources being poured into the teaching of such courses, there is widespread skepticism, backed up by some empirical studies, about their effectiveness, especially in preparing students for genuinely college-level mathematics courses. One should expect more from a quantitative literacy program for undergraduates.

But *is* there an intrinsically "college" part for all students? If agreement can be reached on what "mathematical methods and concepts, their assumptions and implications" every college graduate should understand, does it really matter whether that understanding is acquired before or after matriculation in a college or university? Is it not imaginable that, for example, the goals set for secondary mathematics in the NCTM *Curriculum and Evaluation Standards in School Mathematics* (1989) define an acceptable concept of quantitative literacy? And if so, and if the *Standards* are widely adopted, will there be anything left for the colleges and universities to do in this area beyond supplying suitable remedial experiences for those students who slip through the cracks? To put the matter another way, is it not imaginable that any quantitative literacy appropriately required for a bachelor's degree should in fact be regarded as an appropriate requirement for admission to a college or university?

There are several very large "ifs" in the preceding paragraph. They relate to difficult questions of definition, curricular diversity and inertia, a great lack of homogeneity in the student population, and other inconveniences. A more important consideration, perhaps, relates to the nature of the post-secondary experience.

College students, on the average, are more mature, more experienced, and more thoughtful about their personal goals than they were before they became college students. One does not need to invoke William Perry's scheme to justify a belief that college students should be better able to acquire, and to acquire more deeply, quantitative literacy in any reasonable sense. Indeed, because of the pervasiveness of mathematical ideas in the careers that college graduates usually enter, they should be *expected* to have acquired them more thoroughly and meaningfully than if they had not gone to college.

These ruminations are leading relentlessly to the conclusion that it might be a mistake to speak of "quantitative literacy" as if it were a single, monolithic idea. Surely there are meaningful *degrees* of quantitative literacy, and perhaps it would be useful to identify some of them. Here, we speak of only one—the degree of quantitative literacy appropriately expected of all *college* graduates. As we have suggested, we do not believe that this is identical with the degree of quantitative literacy appropriately expected of all *high school* graduates, even as implied in such a forward-looking statement as the NCTM *Standards.*

Thus we take the stance that, for many reasons, some significant level of quantitative literacy is desirable in all adults; that the amount appropriate for college graduates is greater than that to be expected at the time of graduation from high school; and that the difference is not merely a matter of "remediation."

Cultivation of quantitative literacy at any level is, of course, a matter of teaching and learning. And teaching and learning involve far more than mere identification and communication of appropriate content. There is ample evidence that the traditional "lecture-and-listen" mode of instruction, still probably far more the rule than the exception in American higher education, does not work as well as some other modes—certainly not as well as it should. Particularly for those students who are studying in the mathematical sciences not by their own choice, teaching and learning styles that include active involvement, cooperation, and the personal touch are much to be preferred over those that do not.

So while the emphasis in this report will be on what the elements of quantitative literacy are, we also implore those who are responsible for providing students with classes and other opportunities for developing quantitative literacy to give a great deal of attention to the form those opportunities should take and the manner in which they should be delivered.

What Is Quantitative Literacy?

The Focus Group on Quantitative Literacy was summoned to its conference tasks by words from the National Academy Press' publication *Everybody Counts* (1989):

> Functional literacy in all of its manifestations—verbal, mathematical, scientific, and cultural—provides a common fabric of communication indispensable for modern civilized society. Mathematical literacy is especially crucial because mathematics is the language of science and technology. Discussion of important health and environmental issues (acid rain, waste management, greenhouse effect) is impossible without using the language of mathematics; solutions to these problems will require a public consensus built on the social fabric of literacy.

The conference invitation noted that the issues of quantitative literacy are of critical importance for the lives of all of us, since they shape the future of our democracy as we would like to see it. We are confronted daily with conflicting quantitative information and need to be aware of both the power and limitations of mathematics.

Faced with these weighty statements, conference participants attempted to sift and sort their thoughts into coherent responses to questions posed by the conference moderator. The difficulty of the questions posed was readily acknowledged. One participant who promised a "good answer" soon observed, "The trouble is, I keep changing my answers every few days; this is a tough nut to crack!" Yet another said, "I really am overwhelmed by your questions."

Despite difficult questions, the group uncovered a number of factors that need to be addressed in a quest for quantitative literacy—some arising in a straightforward manner and others arising more subtly.

At first, conference participants struggled with a response to the question of determining the key issues in quantitative literacy for the next five years. The issues were expressed, in one participant's words, in a series of questions:

Is there a valid concept of quantitative literacy which makes sense for all U.S. citizens? If so, how might it be identified? Is it the NCTM *Standards*? If not, what can serve as a guide to educators? A set of options? A set of components which might be mixed in various proportions? What evidence should be considered in answering these questions? If opinions, whose?

Quantitative literacy certainly has different meanings to different people, so in order to work with this concept, some group must say just what we will mean by quantitative literacy. Just what skills, concepts, and behaviors in this general area should students be expected to learn?

Beyond determining a set of objectives (or behaviors) for quantitative literacy, there are attitudinal problems which must be dealt with in our society and in the mathematics community. For many decades it has been "okay" within our culture to dislike mathematics and to seek to "get by" with a level of literacy which would probably have not even been acceptable in a society far less complex than our current one. How can the mathematics community turn around the negative attitudes of our citizens towards quantitative literacy? And how can the resolve of the collegiate mathematics community be heightened to face the enormous remedial task in areas such as problem interpretation and formulation, probability models, and estimation—"areas where students show so little understanding of the basic characteristics of simple operations." It must no longer be acceptable for "educated" citizens to lack quantitative literacy. It must no longer be acceptable for the mathematics community to be willing to see graduated from our colleges and universities students who lack quantitative literacy. But how do we change these attitudes?

The focus group made several suggestions about what quantitative literacy study should include. One idea that was reiterated by several people was that we need to look at concepts in context (i.e., applications areas). "The key is to have the contexts relate to student interest, daily life, and likely work settings." As another put it, "People must be able to absorb quantitative relationships that are expressed in daily life. I see this as essential to an educated citizenry." Among concepts which might be put in context were dimension, distance (or area or volume), model, and data analysis. One participant particularly lauded the study of statistics because it "can generate a need to know basic geometry, algebra, and of course arithmetic." These ideas provide an opportunity to turn negative attitudes around, for they all suggest relevance to the student's life as the student perceives it.

These suggestions immediately pose another problem: how could they be presented in the college curriculum? The mathematics community finds it very difficult to think in any terms other than those courses that have become traditional at the college level. Focus group participants wrestled with thoughts of an "extended" definition of "statistics" and bemoaned the disfavor into which once popular mathematics appreciation courses have fallen. At the college level, new courses must be devised and old courses modified to foster the objectives of quantitative literacy. Courses must be developed on the basis of the determined meaning of quantitative literacy rather than just by asking which existing courses determine quantitative literacy. Just as the Sloan Foundation has conducted some interesting experiments for enticing talented students into mathematics and the Annenberg Foundation supported development of *For All Practical Purposes* to foster a new view of mathematics by the public, so new courses in quantitative literacy must be developed and resource materials made available for their teaching.

The development of new courses is a costly effort, often taking away time from other tasks the mathematician would rather do. Further, producing relevant materials to daily life means materials must, at least in part, be of a "throw away" nature—that is, they will have a shorter life span than text material would often be expected to have. Hence there is need to support the development of a range of both innovative courses that have quantitative literacy as their focus and materials for use in teaching such courses. In addition, once such courses are developed, those that have the greatest success should be disseminated to the broader mathematics community as models. The important question of how quantitative literacy courses relate to other courses in the curriculum will still need to be addressed, of course.

Pre-college Numeracy

Links with pre-college mathematics are also important considerations in the development of quantitative literacy in college students. Driven by efforts to implement the NCTM *Standards*, the pre-college curriculum in mathematics is in a stage of flux. More and more, number sense is being developed in children by consideration of real problems—problems that capture student interest and that arise readily in daily life. One conference participant said, "I think one intent of the new NCTM curriculum is to give students both an appreciation of the power of mathematics and some essential mathematical skills. We can only hope that this curriculum succeeds, but what do we do in the meantime?"

Quantitative literacy at the college level must not wait for the *Standards* to take hold, but it must be approached in a manner that is cognizant and supportive of the changes in the pre-college curriculum which will lead to greater strength in quantitative literacy for everyone.

Another issue to consider is the effect on teacher training programs of the establishment of a quantitative literacy requirement for all college students. Here the recent publication *A Call For Change* may well lead the way to the problem's solution, since it couches its recommendations in terms of *experiences* rather than *courses* a student should have. Surely quantitative literacy courses could offer some valued experiences with mathematics and even some models of ways to teach mathematics.

On the other hand, college professors must right now participate in the teaching of what one participant called "remedial areas"—whether teaching the least or the best-prepared students. Many mathematics instructors do not want to be involved with such teaching—partly because they feel ill-prepared to teach remedial work, but also because they feel students should be quantitatively literate by the time they reach college. In their view those students who are not already quantitatively literate should not be admitted to college. These views lead to another important issue: Is there a difference between quantitative literacy for high school graduates and that for college graduates? If so, what difference in expectations should be made?

One participant felt that the average college-educated person should have a reasonable appreciation for the power and importance of mathematics in our society and culture along with a certain level of mathematical skills.

> I think these two goals are often confused, and that they often work at cross purposes. For example, we make virtually every undergraduate pass a course in college algebra before they graduate (I am not defending this requirement, so boo if you want to). This assures a certain level of

mathematical skills, but these students have absolutely no appreciation of the role of mathematics in contemporary society (or if they do, it is not because of anything we teach them in the mathematics department). In fact, I would wager that the opposite effect is the case: students here regard mathematics as a boring collection of useless tricks for solving dull problems.

On the other hand, another conferee argued that a distinction between high school and college-level quantitative literacy arises, in part, from the form of communication expected: "College implies that one could understand relationships expressed symbolically—whereas high school implies an understanding of everyday level vocabulary." Another suggested that the key word for the distinction between the two levels of schooling is "deeper." "The reasoning and examples can become more sophisticated and the students may allow a broader set of applications at the college level." These comments offer at least two conclusions from the conferees—namely that there is a distinct notion of quantitative literacy for college graduates versus high school graduates, and that algebra does play a role in college-level quantitative literacy.

Establishing Goals

Just what is involved in establishing goals and objectives for quantitative literacy at colleges and universities? At the elementary and secondary school levels, the NCTM *Standards* speak of students acquiring "mathematical power." The term denotes an individual's abilities to explore, conjecture, and reason logically, as well as the ability to use a variety of mathematical methods effectively to solve non-routine problems. This notion is based on the recognition of mathematics "as more than a collection of concepts and skills to be mastered; it includes methods of investigating and reasoning, means of communication, and notions of context. In addition, for each individual, mathematical power involves the development of personal self-confidence." If this standard determines quantitative literacy for high school graduates and students at the college level are required to take a mathematics course in this spirit which goes deeper, will such students be quantitatively literate as college graduates?

Most conference participants argued for some further elements in the baccalaureate experience.

> I also expect quantitative literacy to exhibit a set of behaviors ... (but) all this will fail unless students see that what we are trying to do is really important in their majors.

> On my campus we are gearing up for 'writing across the curriculum' ... and I contend that one can simply replace 'writing' with 'mathematics' in virtually everything I have read about writing across the curriculum and get a true statement.

> To me mathematics is a very broad topic ... I see mathematics in politics, sociology, history, etc.

> I think there is a general move to writing in mathematics—as well as mathematics across the curriculum.

Thus the participants argued not only for a mathematics course that fosters quantitative literacy for college students, but also that quantitative literacy goals be connected with the broader curriculum and also involve writing.

Two other connections with the broader curriculum surfaced in the discussions. Although they are not new, in the current context they elicit deeper thought. These are "helping people to be life-long learners" and the issue of "appreciation" of mathematics. Regarding life-long learning and quantitative literacy, one conferee said:

> I'm not sure how one goes about doing this, but part of it must involve inciting curiosity about numerical phenomena. I think people tend to be curious about numerical things (what are my chances of winning the lottery?), but think the solutions are somehow mysterious and beyond them. Thus, we need to provide simple ideas for handling numbers that do not involve messy algorithms and can be seen as widely applicable building blocks to number sense (a jackknife, not a surgeon's scalpel?).

As for "appreciation" of mathematics, it was argued, hopefully, that appreciation would come with the acquisition of skills (where skills are intended in the broadest sense as terminology). Another voiced the opinion that students might profit from the study of some mathematical models: "Perhaps students can learn to appreciate the power of mathematics by understanding the simpler components well enough to envision the role of a more detailed model, even if they have no interest in constructing the model themselves."

Yet another connection which the group brought up is the role that the ability to use hand calculators and personal computers should play in quantitative literacy. One conferee exclaimed, "I must share my support for a hands-on technologically enriched curriculum in which a person becomes quantitatively literate—a person *becomes*." Technology is used, among other things, to clarify, to depict, and to manipulate data; implied in its use are the needs for estimation skills and skills at analysis of error. Quantitative literacy must include the effective use of common machines in order to bring about increased understanding of mathematical situations and to solve problems either precisely or in an approximate form. Since such use beyond the sciences and mathematics is still in a developmental stage, this idea requires special wisdom on the part of those who teach with calculators or computers. College teachers in every discipline need ideas and experience in how to use these machines appropriately and well.

Research

Certainly whatever efforts are exerted to make college students and society more quantitatively literate must be handled so as to see that those groups that have been under-represented among our educated citizenry are not treated adversely. Research needs to be done to see how quantitative literacy programs impact not only the attainment of the desired education, but also to see if various groups of people are being affected differently, or if different strategies must be adopted for certain groups of learners based on learning styles. If an important part of quantitative literacy programs is to be study-in-context of problem situations, the contexts must not be selected so as to be understood only by a subgroup of the students in a program. In the past, women and minority groups may have lacked the "key to opportunity" which being quantitatively literate provides, but this must not continue inadvertently in a design of new programs for quantitative literacy for all college graduates.

Implied in the call for research in the previous paragraph is a strategy of assessment. Colleges need to know where their students are on the road to "becoming" quantitatively literate in order to know what the students must yet attain. Some portion of the assessment by the National Center for Education Statistics on "The State of Mathematics Achievement" might be helpful in establishing baseline information for aspects of quantitative literacy. (See Appendix A for a more thorough discussion of an appropriate research data base.) Colleges or universities might also consider some entrance level assessment in quantitative literacy—a task which may serve a dual role of information for the school to use in advising the individual student and of political force. As one conferee said, "...entrance level assessment in quantitative literacy at the university level would help impress on high school students a certain minimal competency level."

However, beyond these thoughts the conference members did not come to grips with means for assessing whether the goals for quantitative literacy have been achieved for an individual student or whether it is desirable to have a "rising junior" examination as in the state of Florida where a student may not become a junior without exhibiting some minimal skills in mathematical thinking on a standardized test. Perhaps some new instruments are needed to truly validate levels of quantitative literacy, especially when consideration is given to the goals and objectives which the group discussed as being part of quantitative literacy.

If at the college level quantitative literacy includes the establishment of connections between mathematics and other disciplines or is fostered by mathematics across the curriculum, then mathematical sciences faculty members must reach out to colleagues in other departments on campus and to administrators to effect change which will enable students to become quantitatively literate. Ideas on how to make the connections need to be developed and disseminated. (The only group of courses developed so far of this nature appears to be those courses on quantitative reasoning put together with the support of the Sloan Foundation in the New Liberal Arts Program; a description of one such course, from Mount Holyoke College, appears in Appendix C.)

Outreach

Along the lines of what is working, it should be pointed out that the NCTM *Standards* have alerted the teaching population at the elementary and secondary levels to the need for change in the teaching of mathematics and serve in part as a model for change. The NSF project on statistics which was organized through a joint committee between the NCTM and the American Statistical Association "with the goal of providing curriculum materials and in-service training so that mathematics teachers in the secondary schools could effectively incorporate basic concepts of statistics and probability into their teaching of mathematics" has met with outstanding success. These have given excellent support for cooperative learning, provided a basis for writing in a mathematics class, and engendered positive attitudes by students toward mathematics. Workshops and courses have been taught so as to model how these materials should be used. Similar kinds of activities will likely be necessary at the college level, although materials need not all be statistical in nature. (See Appendix B for examples of other related course materials.)

Besides students and colleagues in our colleges and universities, who else do we need to reach to effect change in the area of quantitative literacy? Groups advanced were: the press, parents, school boards, PTA's, politicians, and CEO's. One participant observed that many states have state mathematics coalitions supported through federal funds and the Mathematical Sciences Education Board. The people we want to reach to bring about change should be represented on these coalitions. "A plan could be developed that works with the mathematics coalitions of every state to begin to change the quantitative literacy level of all people."

Part of such a plan might be an idea advanced by another conferee who felt that we had to try to educate parents so they and their student children can work together on improvement of reasoning skills. He noted, "Two projects in this direction are the *Math Kits* program of the MSEB and the recent *Math Power* series for middle schools from AAAS, one of which is entitled *Math Power at Home*. These publications are at the school level, but I think we should understand and improve that level so that we might achieve some carry-over to the college level."

Further parts of a plan must be to acknowledge that politicians are the decision-makers in our society; they must have some quantitative literacy skills of their own and understand the needs for education in quantitative literacy if they are to support funding for programs in mathematics education. It was also noted by one participant that "my experiences suggest that politicians have relatively low regard for educators but high regard for successful business people," so "we must use business and industry connections" to reach them.

Another idea advanced regarding how to reach the public is one which could also have the effect of heightening the awareness at colleges and universities of the need for improved quantitative literacy.

> We as educators, and our students, should be more forceful in making the media do a better job of handling quantitative information, and calling them on it when they do a poor job. Critical reading of such articles is one of the things we should teach, just as English teaches critical reading from the standpoint of grammar and style. We should simply not stand for some of the things we see in advertising, or the misuse of polls, or the incorrect interpretation of medical studies. ...We must show people the importance and correct interpretation of good uses of mathematics (the Physician's Health Study, tracking the CPI and unemployment rates), and point out the problems and disasters that occur from no or incorrect mathematical modelling (the Challenger disaster, the flap over AIDS testing).

The College Algebra Debate

In an editorial in the Washington *Post* on April 20, 1991, Colman McCarthy gave his response to "Who Needs Algebra?"

> Would millions of high school students trudge into their algebra classes if it weren't a gate through which they were forced to pass to enter college? And in a few years would they submit to college algebra if it weren't a requirement to graduate? Not likely. Algebra is more loathed than learned, more endured than embraced. It is more memorized to pass tests than understood to comprehend problems.

Is McCarthy right? Many colleges and universities require college algebra of their graduates, but why do they do it? There are a number of factors involved, at least one of which is that it is said to assure that students have had the type of intellectual challenge which the study of college mathematics represents. But there are other reasons too.

College algebra fits neatly with the standard curriculum in high school and with the lower division courses taught in college which are required in majors outside mathematics. Placement tests can be easily written, administered, and evaluated to see if a student needs college algebra. Proficiency examinations (or CLEP tests) are also easy to give and evaluate. Further, mastery of the computational skills of college algebra is essential for continued study in mathematics for most students. Thus, if students take college algebra upon entering college, they may leave open choices for majors which require additional mathematics. In an editorial in a recent issue of the *UMAP Journal* (Vol. 11, No. 4, 1990), Joe Malkevitch argues that preserving access to study of mathematics, science, and engineering remains *the* reason students are pressured into studying algebra.

While it is true that mathematics is the "language of science," it is increasingly true that social and management sciences are also using algebraic techniques and computational skills. However, other disciplines are using much mathematics that is not algebra as well.

At most colleges and universities the focus of study in a college algebra course is on those computational skills needed to do calculus (although the course is also used as prerequisite material for courses in finite mathematics, statistics, etc.). With this focus exclusively, college algebra does not make a good general education course—a course which should convey how mathematicians think and develop their ideas. Further, a college algebra course with this focus does *not* meet the objectives of quantitative literacy as suggested by our e-mail conference discussion.

Many in the mathematics community view the standard college algebra course as "in trouble." In an article "What's Wrong with College Algebra?" appearing in the *UMAP Journal* (Vol. 12, No. 2, 1991), Peter Lindstrom lists eight areas of disagreement among mathematics teachers about the course. Among these areas are three which were brought out in a debate titled "Resolved—All College Graduates Should Know College Algebra" that was sponsored by the CUPM Subcommittee on Quantitative Literacy at the January 1991 AMS/MAA meeting in San Francisco. First, many students taking the course are not prepared for it (so the course is often compromised). Secondly, a college algebra course which does not focus on computational skills tries to do too many things to succeed in conveying essential skills for the calculus or other courses for which it is prerequisite. And thirdly, a course which stresses problem solving (not routine problems and skills development) is a more appropriate course for *all* college graduates. Implied in the discussion was the thought that a modified college algebra course which stresses problem solving with problems connected with daily life and which uses hand calculators as a natural tool in problem solving may serve neither of the two objectives it seeks to meet.

The proper balance of elementary algebra and problem solving is yet to be determined in the composition of appropriate quantitative literacy for all college students. Yet to be determined also is the effective use of hand calculators in the teaching, learning, and use of college algebra taught as prerequisite to the further study of mathematics. But now is the time to determine both.

Challenges

These latter thoughts show what consequences can occur, and may occur more and more, if the key issues in quantitative literacy are not addressed. Quantitative literacy is essential as a foundation for democracy in a technological age. Without it our citizens are ill-equipped to make the responsible decisions a democracy places in their hands, and we can expect more and more disasters—if not extinction from poor management of the environment. Citizens need to be able to understand research results which are often expressed only in quantitative terms.

The time has come for professional groups such as the MAA to speak out on what quantitative literacy for college students should mean. As one participant put it, "Part of the reason, in my opinion, that a state ... would try to define quantitative literacy as a college algebra course is because the mathematics professional groups are leaving such a void there." Vague generalities are not good guidelines, and good guidelines are greatly needed as to what quantitative literacy entails and how it can be accomplished or its level improved. Such guidelines could be helpful to faculty and deans of colleges as well as to others.

There is also need for research. As one participant said, "If we are to improve teaching and learning, we must know something about how people learn, especially about their intuitive understanding of the world." The more we understand about how students learn those skills, concepts, and behaviors we choose to call quantitative literacy, the better we should be able to teach them.

These were the issues the Focus Group on Quantitative Literacy discussed. Interestingly, missing from the discussion was any mention of "math anxiety" or "mathophobia." Participants seemed to be following the thought pattern of what should a college educated person be able to do. Quantitative literacy was not viewed as a spectator sport in these discussions. It was a vital need for everyone!

Focus Group Participants

mabernha@ecuvm1.bitnet	ROBERT BERNHARDT, *East Carolina University.*
bushaw@wsuvm1.bitnet	DONALD BUSHAW, *Washington State University.*
jad@scsla.howard.edu	JAMES DONALDSON, *Howard University.*
0003332959@mcimail.com	SOLOMON GARFUNKEL, *COMAP, Inc..*
calculus@math.harvard.edu	DEBORAH HUGHES HALLET, *Harvard University.*
matccn@depaul.bitnet	CAROLYN NARISIMHAN, *DePaul University.*
rn02@swtexas.bitnet	ROBERT NORTHCUTT, *Southwest Texas State.*
ga0181@siucvmb.bitnet	KATHERINE PEDERSEN, *Southern Illinois University.*
jjp@euclid.math.purdue.edu	JUSTIN PRICE, *Purdue University.*
sally@zaphod.uchicago.edu	PAUL SALLY, JR., *University of Chicago.*
scheaffe@orca.stat.ufl.edu	RICHARD SCHEAFFER, *University of Florida.*
jsolomon@math.msstate.edu	JIMMY SOLOMON, *Mississippi State University.*
sons@math.niu.edu	LINDA SONS, Moderator, *Northern Illinois University.*
jtruxal@sbccmail.bitnet	JOHN TRUXAL, *State Univ. of New York at Stony Brook.*

References

1. Committee on the Mathematical Education of Teachers. *A Call for Change: Recommendations for the Mathematical Preparation of Teachers of Mathematics.* Washington, DC: Mathematical Association of America, 1991.

2. *Entry-level Undergraduate Courses in Science, Mathematics, and Engineering: An Investment in Human Resources.* New Haven, CT: Sigma Xi, the Scientific Research Society, 1991.

3. *For All Practical Purposes: Introduction to Contemporary Mathematics.* New York: W.H. Freeman and Company, 1988, 1991.

4. *Math Power at Home.* Washington, DC: American Association for the Advancement of Science, 1991.

5. National Center for Educational Statistics. *The State of Mathematics Achievement.* Washington, DC: U.S. Department of Education, 1991.

6. National Commission on Excellence in Education. *A Nation At Risk: The Imperative for Educational Reform.* Washington, DC: U.S. Government Printing Office, 1983.

7. National Council of Teachers of Mathematics. *Curriculum and Evaluation Standards for School Mathematics.* Reston, VA: National Council of Teachers of Mathematics, 1989.

8. National Governors' Association. *Results in Education 1987: The Governors' 1991 Report on Education.* Washington, DC: National Governors' Association, 1987.

9. National Research Council. *Everybody Counts: A Report to the Nation on the Future of Mathematics Education.* Washington, DC: National Academy Press, 1989.

10. *Quantitative Literacy Series.* (A project of the American Statistical Association.) Palo Alto, CA: Dale Seymour Publications, 1987.

11. Sons, Linda R. "What Mathematics Should Every College Graduate Know?" *UME Trends*, October 1990.

12. Steen, Lynn Arthur (Ed.). *On the Shoulders of Giants: New Approaches to Numeracy.* Washington, DC: National Academy Press, 1990.

13. "The New Liberal Arts Program: A 1990 Report." Sam Goldberg (Ed.). New York: Alfred P. Sloan Foundation, 1990.

Appendix A: Assessment and a Research Data Base

For many years the National Assessment of Educational Progress (NAEP) has conducted a mathematics assessment of fourth, eighth, and twelfth graders. The assessment is designed to measure mathematics proficiency in six content areas: (1) numbers and operations; (2) estimation; (3) measurement; (4) geometry; (5) data analysis, statistics, and probability; and (6) algebra and functions. A new aspect of the 1990 program was that states could participate on a voluntary basis in assessment of eighth graders enabling state comparisons with each other and the nation. Thus the U.S. Department of Education issues "The State of Mathematics Achievement: NAEP's 1990 Assessment of the Nation and the Trial Assessment of the States (Executive Summary)" as well as for those participating states a separate report available through that state's Department of Education.

The purposes of the NAEP data gathering are expressed as: "...to provide a detailed portrait that can be used in examining where the nation is in relation to its overarching goals for mathematics education, and how far mathematics educators have moved toward meeting their standards ...This information can be used to monitor students' progress in achieving what has been recommended for reform in school mathematics, to explore issues of equity in opportunity to learn mathematics, and to examine both school and home contexts for educational support."

A perhaps lesser-known study is the Longitudinal Study of American Youth (LSAY) initiated in 1986 by a grant from the National Science Foundation. The primary objectives of the LSAY were and continue to be improvement in our understanding of:

• The development of student skills in science and mathematics.
• The role of student attitudes toward science and mathematics in the development of skills in those areas.

- The development of a student preference for or against a career in science, mathematics, engineering, and medicine.
- The influence of parents and other family members on the development of student attitudes toward and skills in science and mathematics and the emergence of career preferences.
- The influence of school curriculum and environment on the development of student attitudes toward and skills in science and mathematics and the emergence of career preferences.
- The influence of teacher training and classroom practice on the development of student attitudes toward and skills in science and mathematics and the emergence of career preferences.
- The influence of informal science education on the development of student attitudes toward and skills in science and mathematics and the emergence of career preferences.
- The influence of scientific literacy on the development of a sense of citizenship efficacy in regard to public policy issues involving science and technology.

The LSAY follows students through the years and has mined some models on predicting mathematics achievement, effects of ability grouping in middle school science and mathematics on student achievement, and more. For example, analysis of young adult data showed that the extent of formal education in science and mathematics is the most important influence on attentiveness to and interest in science. The research data base generated by this study should enable some researchers to obtain a better understanding of how students learn. Data is collected so as to trace students into their college years.

The LSAY has been directed by Jon Miller and Robert W. Suchner of the Public Opinion Laboratory and Social Science Research Institute at Northern Illinois University, DeKalb, Illinois. Available now is the LSAY Codebook: *Student, Parent, and Teacher Data for Cohort One for Longitudinal Years One, Two, and Three (1987-1990), Volume 1* (March 1991). Some reports of work with the data have been given at meetings of the American Educational Research Association. Among these are "Some Models to Predict Mathematics Achievement," and "Modeling NAEP Items by Cognitive Process" by Robert W. Suchner, and "The Longitudinal Study of American Youth and the National Education Longitudinal Study of 1988: A Comparison" by Thomas B. Hoffer.

Appendix B: Course Materials

Billstein, Rick and Lott, Johnny W. *Mathematics for Liberal Arts: A Problem Solving Approach.* Benjamin Cummings Publishing Company, 1986.

This book involves a survey of some elementary mathematics—topics which have been standard in mathematics appreciation courses—but it presents and uses the heuristics of problem solving as an integral part of the mathematics. The book does not presuppose specific high school mathematics. Problems emphasizing calculator usage are included in problem sets, and "Computer Corners" present programs in BASIC and Logo.

Caruth, J. Harvey. *Algebraic Reasoning Motivated by Actual Problems in Personal Finance.* The University of Tennessee, Knoxville, 1990.

"Traditional treatments of most mathematical concepts covered in this text may be found in virtually any high school or college algebra text." The book features actual problems in personal finance to motivate algebraic concepts—problems relevant to most college students' experience such as borrowing money for the first year of college, or paying for a car or house with monthly payments. Emphasis is placed on reasonableness of results and the development of "number sense." The use of calculators and computers is included in text treatment of problem situations.

For All Practical Purposes: Introduction to Contemporary Mathematics, Second Edition. W.H. Freeman and Company, 1991.

This book is a revised version of the book written by some 14 authors and edited by Lynn Steen. The original version was a project of COMAP, Inc., and was intended to be used in conjunction with the telecourse available for PBS through funding by the Annenberg Foundation. The book is intended for a one-term course in liberal arts mathematics or a course that surveys mathematical ideas. Presupposes some ability in arithmetic, geometry, and elementary algebra. Emphasis is placed on connections between contemporary mathematics and modern society rather than on the capacity of the student to *do* mathematics.

Growney, Joanne Simpson. *Mathematics in Daily Life: Making Decisions and Solving Problems.* McGraw-Hill Book Company, 1986.

This book is intended for use in college or university general education mathematics courses that are focused on developing abilities in quantitative and logical reasoning. It presupposes and emphasizes mainly arithmetic in decision making and problem solving. Use of mathematics (not the mathematical topics) is stressed. The text, its topics, and its exercises are quite nonroutine in nature. Students are expected to use hand calculators in completing exercises.

Pollatsek, Harriet and Schwartz, Robert. "Case Studies in Quantitative Reasoning: An Interdisciplinary Course." Extended Syllabi Series of the New Liberal Arts Program, Alfred P. Sloan Foundation, 1990.

This is a syllabus for a course developed at Mount Holyoke College which is open to all students at that college (as a general education course). The course teaches quantitative methods in the context of how they are used. A Macintosh computing facility is available for student use. The goals of the course are "to help students strengthen their analytical skills and acquire a more confident understanding of the meaning of numbers, graphs, and the other quantitative materials that they will encounter in many subsequent courses." The three main units are: I. Narrative and Numbers: Salem Village Witchcraft; II. Measurement and Prediction: SAT Scores and GPA; III. Rates of Change: Modeling Population and Resources.

Schwartz, Richard H. *Mathematics and Global Survival, Second Edition.* Ginn Press, 1990.

This book, "written to make students aware of critical issues facing the world today, and to help them respond effectively," might be termed a developmental mathematics book—students are only assumed to have elementary computational skills. Mathematical concepts used are some arithmetic ideas (percents, ratios, etc.) and some ideas in elementary probability and statistics. Almost every problem is related to an issue of global survival.

Sons, Linda R. and Nicholls, Peter J. *Mathematical Thinking in a Quantitative World.* Preliminary Edition. Kendall/Hunt Publishing Company, 1990.

This book is written for a college course in quantitative reasoning which presupposes two years of college preparatory high school mathematics including one year of high school algebra. The intent of the material is to help "develop in the student a competency in problem solving and analysis which is helpful in personal decision-making; in evaluating concerns in the community, state, and nation; in setting and achieving career goals; and in continued learning." It is assumed students will use elementary hand calculators. Emphasis is placed on *use* of the mathematics students have studied through facing them with problem-solving situations which are relevant to their daily lives. Both text and exercises seek to provide the bridge students need from their previous more computational-skills oriented mathematical experiences to higher levels of mathematical thinking.

Wattenberg, Frank. *Personal Mathematics and Computing: Tools for the Liberal Arts.* McGraw-Hill Publishing Company, 1990.

> This book is intended as a computer literacy tool but also as a tool to teach a student to use mathematics to reason about a variety of important real problems. The text teaches True BASIC programming but emphasizes applications rather than mathematics and programming. It presupposes college preparatory algebra and geometry. Chapter topics are probability and statistics, economic models, optics, local aid distribution, and population models. The book is a text for one of the courses in the New Liberal Arts Program which was developed through funding by the Sloan Foundation.

Appendix C: Case Studies in Quantitative Reasoning

by Harriet Pollatsek and *Robert Schwartz,* MOUNT HOLYOKE COLLEGE

Case Studies in Quantitative Reasoning teaches quantitative methods in the context of how they are used. The course is distinctive in a number of ways. For instance, it has been designed and is taught by members of many departments: Biological Sciences, Economics, History, Mathematics, Statistics and Computation, Physics, Psychology, and Education. The course has been developed in discussions among faculty over the past few years. Its structure differs from that of a "regular mathematics course" in that it includes lectures, labs, and small discussion sections (of about fifteen students). A Macintosh computer facility has been created especially for Quantitative Reasoning (QR) students. The Macs are used for writing papers, doing data analyses, making graphical displays, and creating models of changing systems.

The primary difference, though, is one of approach. The course, which has no prerequisites, is designed to appeal to students with a broad range of academic interests and widely differing mathematical backgrounds—from math-phobes to calculus-philes. It is not a simple presentation of technical methods followed by practice problems. Instead, case studies from a variety of disciplines form the subject matter of the course. Different quantitative methods are introduced and used in the attempt to develop understanding of these examples. The emphasis is not on rote computation, but on reasoning; not on formulas, but on ways to construct and evaluate arguments. The goals are to help students strengthen their analytical skills and acquire a more confident understanding of the meaning of numbers, graphs, and the other quantitative materials that they will encounter in many subsequent courses, no matter what their majors.

Each semester three sections are offered with room for about 45 students. Distribution credit in mathematics is given for successful completion of the course. However, it is recommended that students work for a second semester to reinforce and extend their mastery of quantitative arguments. Continuations that particularly emphasize quantitative reasoning skills include an interdisciplinary Topics in Quantitative Reasoning, Elementary Data Analysis and Experimental Design, Psychology 201 (Statistics), and Calculus I.

Here is an outline of the three major units of the course.

I. NARRATIVE AND NUMBERS: SALEM VILLAGE WITCHCRAFT

Witchcraft in seventeenth-century New England forms the central problem for investigation. The major project is to write a paper formulating and discussing a hypothesis about the relationship between wealth and power as reflected in the historical records for Salem Village during the 17th century.

This first section concentrates on what can be called "exploratory data analysis," that is, the search for meaningful patterns in numerical data. There is heavy emphasis on graphical and other methods as tools for finding and presenting patterns. The overall goal is to facilitate the assignment of meaning to a set of data, stressing the process of translation between quantitative patterns and plausible explanations. Hypothesis testing is one method studied to safeguard against building up a theory on the basis of numerical coincidence or mere chance.

Skills:	Concepts:
contingency tables	cross-classification
percents	standardizing comparisons
bar graphs	defining meaningful categories
mean, median	choosing variables
percentiles	constructing an argument
expected value	simulation
chi-square	probability as a ratio (relative frequency)
using statistical software	bias
probability as a measure of surprise	logic of hypothesis testing
	null and alternate hypotheses

II. MEASUREMENT AND PREDICTION: SAT SCORES AND GRP

Aptitude and achievement test scores, grade point average, choice of major and other background information for recent Mount Holyoke graduates form a rich data set for investigation. Students formulate and test hypotheses and study the relationships between hypotheses, formal models, predictions, and actual results.

This section concentrates on hypothesis testing, experimental design, and modelling. The ideas and techniques of measuring differences, the identification of critical and missing variables, and the role of disconfirming evidence in the construction of an argument are emphasized. Important statistical concepts such as scales of measurement, distributions, averages, variation, correlation, test reliability, and test validity are introduced.

All skills and concepts from Unit I are reinforced, plus:

Skills:	Concepts:
xy plots	measurement
standard deviation	variability
equation of line	correlation: strength of relationship
slope	correlation: spread about best-fitting line
correlation	prediction
best fitting line	$(correlation)^2$: predictive advantage
$(correlation)^2$	correlation vs. causation
linear models	missing variables and confounding factors
	correlation and hypothesis testing
	reliability and validity

III. RATES OF CHANGE: MODELLING POPULATION AND RESOURCES

The rate of change of a quantity is the focus of this section. Here case studies are drawn from the sciences, particularly ecology. The effect on a population of birth and death rates, the spread of disease, and the decay of radioactive waste are but a few of the examples in which rate of change is a natural quantity to measure. Laboratories help develop the concept of rate of change by modelling changing systems over time.

Rate of change is a concept that is usually covered in calculus courses, but its usefulness in understanding a variety of situations makes it a natural topic of study. As in the two earlier sections of the course, the formulation of hypotheses, the translation of data into argument, and the construction of models are used to further understanding.

Skills:	Concepts:
slope as rate of change	rates of change
arithmetic growth	linear models
geometric growth	exponential models
net change	limited growth (logistic)
graph reading	rate equations
simple algebra	mathematical modelling
using *Stella*	simulation

The Laboratory

The laboratory is a crucial element in the teaching of Case Studies in Quantitative Reasoning. Each week students attend a three-hour computer laboratory in addition to the lecture and two discussions. A special laboratory equipped with Macintosh Pluses is set aside for their use. Laboratories are taught (and laboratory reports are graded) by a laboratory instructor. The laboratory instructor also holds some office hours (supplementing faculty office hours), participates in weekly staff meetings, and attends lectures and discussions. After many variants, this arrangement has proved to be very satisfactory. Students have enough time and support to do the work we expect of them, the load on faculty teaching the course is manageable, and the laboratory instructor gives continuity as the faculty involvement changes from semester to semester.

The Macintoshes were chosen because they are extremely user-friendly and also because excellent software is available for them. For the first two units of the course a spreadsheet-style statistical package called *StatView 512+* by Abacus Concepts, Inc. is used. Its attractive features include the ability to handle both categorical and numerical variables, to permit the use of meaningful verbal labels for variables, and to create varied visual displays of data.

For the third unit, students use *Stella* by High Performance Systems. By the use of clever visual icons, *Stella* permits the student to model dynamical systems of considerable complexity without the explicit use of calculus. The central visual metaphor is plumbing: quantities that grow or shrink appear as the contents of "bathtubs." "Pipes" feed into and out of the tubs, each with a "valve" which determines the rate of increase or decrease. These rates may, in turn, depend on other quantities in the system. For example, a population's birth rate (the valve on the in-flow pipe) may be a constant (arithmetic growth) or may be a constant multiple of the current size of the population (geometric growth). In the latter case, an arrow is drawn from the population tub to the valve, and the valve's flow rate may then be defined as the desired constant times the population size.

The typical pattern in each unit is that the first two laboratory assignments are highly structured and require formal laboratory reports. The third laboratory assignment is a somewhat less structured preparation for that unit's paper. An informal written report on the third lab—in effect, a very rough draft of the paper—is brought to the faculty instructor for a conference before the final version of the paper is prepared. In the last week of the unit, the laboratory is simply open for students to drop in as needed. Only the first unit deviates from this pattern; it lasts five weeks, so the first laboratory is an introduction to the machines themselves.

Over the three years since the course was first offered, more and more students have had some prior computer experience. However, many students still begin with no experience—and often a great fear of computers. Even these most fearful students become confident computer users by the courses' end.

Appendix D: Voices from the Focus Group Conversation

You pose extremely difficult questions. First, let me emphasize that I do not think Sloan has found the answers or even acceptable answers. Sloan has generated some interesting experiments (Mount Holyoke, Wellesley, Grinnell, for example), but these are designed for talented students who need to be enticed into mathematics. The examples are probably not useful with average college students (or those below average) who dislike math.

While the key problem is, of course, in the pre-college arena, we in colleges must face the remedial task in areas such as problem interpretation and formulation, probability models, and estimation—areas where students show so little understanding of the basic characteristics of simple operations.

—John Truxal, SUNY at Stony Brook

I need to address ideas within a context—I guess I am a "situationalist." Thus, I have to define the situation from which I am responding before I respond. For example,

1. What would be outcomes of schooling (K–4, 5–8, 9–12, 13+) in the area of quantitative literacy?
2. What populations would we be addressing?

As scientists, we are used to responding to "data" (i.e., facts) but here we don't have them. What is known about quantitative literacy at this time? What does research tell us in this area? For example, is it reasonable to assume that the general public perceives quantitative literacy to be the province of the select few? Do we have any measurement on quantitative skills as they are interpreted by the general public?

In order to formulate needs to be addressed over the next five years we need to know where we are relative to quantitative literacy. Thus, my first "need" for the next five years is a well-defined research program that describes where we are.

To continue in this vein, we need an oversight committee that endures for sufficient time to gather data and to make sure that the research is helping to define where we are as a quantitatively literate nation. As I continue on this thought, I realize that in order to define the research needed, we need some commonly accepted definition of quantitative literacy. Thus, we need a definition of quantitative literacy, or at least some measurable entities associated with this concept.

So now I need to describe the behaviors that I would expect of a quantitatively literate person—or society. From these behaviors we can decide whether or not people have them, we can formulate means to "educate" people to have them, and we will be on our way to a five-year (or 500-year?) plan.

1. What is the effect of technology on proliferation of data?
2. What is the effect of technology on understanding concepts involved in data analysis?
3. How can we educate our population to the sophistication needed to understand decision making in the context of uncertainty?
4. How much is ignorance of quantitative skills a matter of beliefs?
5. What would be a core curriculum for quantitative literacy, K–13+?
6. How could quantitative literacy concepts be used as the environment in which to teach (re-teach?, develop?) arithmetic skills and concepts?
7. How can we educate the public to understand "modelling" as a legitimate problem-solving technique?
8. What computer technology will be needed to answer the above questions?
9. What mathematics research must be supported to answer the above questions?
10. Are women and minorities affected (differently?) by lack of quantitative literacy skills?
11. What effect does a decision relative to quantitative literacy have on the preparation of all teachers? of mathematics teachers? of elementary teachers ? of secondary teachers? of university teachers?
12. What research agenda do we set up to answer these questions?

These are my questions—from all contexts!

—Katherine Pedersen, Southern Illinois University

On my campus, we are gearing up for "writing across the curriculum," so that is on my mind. And I contend that one can simply replace "writing" with "math" in virtually everything I have read about writing across the curriculum, and get a true statement.

—Robert Bernhardt, East Carolina University

It seems to me that it would be more useful to decide *what we want to mean* by quantitative literacy than to try to discover *what is meant* by it, e.g., in comparison with mathematical literacy. The main question, I think, is: what things (skills, concepts, behaviors) in this general area should students be expected to learn? The total answer might serve as an "extensive" definition of quantitative literacy. After we have that answer, it will be time to talk about courses, curricula, etc.

—Donald Bushaw, Washington State University

My feeling is that we need to look at "concepts in context." Contexts, i.e., application areas, will change with time and fashion. Today, we might do space, robotics, etc., but tomorrow (even five years from now) this will change. The key is to have the contexts relate to student interest, daily life, and likely work settings. The concepts, likely, will have more staying power. Undoubtedly we'd all have different lists, but mine would include: chance, model, cardinality, dimension, distance (area and volume), and data, among others. I find this "concepts in context" a useful way to approach quantitative literacy.

—*Sol Garfunkel, COMAP*

I see two problems concerning quantitative literacy:

1. That the average college-educated person have a "reasonable" appreciation for the power and importance of mathematics in our society and culture; and
2. That the average college-educated person have a certain level of mathematical skills.

I think these two goals are often confused, and often work at cross purposes.

—*Robert Bernhardt*

I do feel that there is some "truth" in defining quantitative literacy as understanding and communicating relationships—and other such words. Some questions and comments:

1. Are we discussing quantitative literacy or mathematical literacy?
2. Are we creating a mathematics course? Is this a means rather than an end?
3. To me quantitative literacy and mathematical literacy are *not* the same.
4. If we are looking at how to develop quantitative literacy in a classroom situation then we are looking at technology and communication. We are looking at data collection and analysis—maybe even more than "statistics." We are looking at *all* estimation skills—we are looking at communication of data and interpretation of this data, making inferences, justifying results, communicating results—and using technology to clarify, depict, manipulate, etc.
5. There is no question that deciding whether or not information communicated by quantitative expressions is reasonable and relevant is an important part of quantitative literacy.
6. I must share my support for a hands-on technologically enriched curriculum in which a person *becomes* quantitatively literate.
7. Must we not, however, specify some behaviors of the person who is quantitatively literate?

—*Katherine Pedersen*

I begin with a little history of my involvement with quantitative literacy issues—all from the perspective of a statistician, not a mathematician.

I have been involved with an NSF-funded project called Quantitative Literacy for about ten years now, and I thought it interesting that this Focus Group has the same title. Our quantitative literacy project was organized through a joint committee between NCTM and the American Statistical Association (ASA) with the goal of providing curriculum materials and in-service training so that mathematics teachers in the secondary schools could effectively incorporate basic concepts of statistics and probability into their teaching of mathematics. The project has been successful beyond our wildest imaginings and formed much of the background for the NCTM *Standards* committees as they wrestled with the problem of how to enhance mathematics education. Thus the *Standards* does, indeed, reflect a data analysis perspective not only in the statistics strand but also in other strands such as algebra, measurement, and functions. By beginning with a real problem of interest to the students, collecting data pertinent to the problem, and then analyzing the data as objectively as possible, one can motivate and illustrate many of the basic concepts in secondary school mathematics. (On a more creative day, not a Monday morning, I might say "all of the basic concepts.") The numbers with which students work must have a context (they then become data) which they can understand and view as relevant to their lives.

You might be interested in our working definition of quantitative literacy: *A complex of skills and knowledge associated with the collection, display, and analysis of data.* This definition is much broader than what most would take as a definition of statistics, but it attempts to include all of the ways that students may interact with data throughout their lives.

I might add, at this point, that American industry has finally caught on to the importance of statistical process control to maintain and improve quality and productivity. SPC methods use elementary data collection and analysis techniques much in the spirit of quantitative literacy. Industries teach these techniques to all levels of employees by in-house courses that involve much hands-on work with real data. The techniques are largely graphical and are used by many with poor mathematics backgrounds. So, I think we have something to learn from the experiences of industry and can, in turn, develop some ideas that would be beneficial to the long term goals of industry. Many industries would like to find ways to work cooperatively with colleges and universities.

So, what about my broader thoughts on quantitative literacy? To be literate means to be able to read and write, to understand what others are trying to convey, and to convey information to others. Thus, quantitative literacy must mean to be able to understand what others are attempting to convey in numerical form and to convey information to others in numerical form. I used the term "numerical form" rather than "mathematical form" intentionally. I think our basic goal is to build an understanding of numerical information—number sense, if you will—and not an understanding of abstract symbolism. The latter would be nice, but is a luxury at this stage.

To combine the ideas expressed above, I think we can and should teach people "number sense" by emphasizing applications with real data that come from their own experiences in the world around them. This approach makes use of mathematics, but in the context of application—real application. The goal is to help people to think correctly about numbers and use data to help make intelligent decisions in life. All of standard secondary school mathematics can be incorporated here, as well as most introductory college mathematics.

Arithmetic: Constructing and interpreting data tables involves fractions, percent, ratio and proportion, averages of different types, etc. Also, estimation is important to the understanding of computations. (Did my calculator give me the correct value of the mean? Is this the mean I want?)

Algebra: I view algebra as an understanding of relationships, most of which can be understood without memorization of an algorithm or appeal to abstract notation. I know some college students who cannot figure out the gasoline mileage they are getting on their cars. I once had a secretary who often asked me something like the following: "If the fringe benefits rate is 25% and we are paid a total of $10,000 for the summer, how much of that is actually salary?" I believe students can be taught how to reason through such problems by concentrating on experiences with real data. So what if ten different students reason out the solution in ten different ways?

Geometry: Many of the measurement problems with which we are confronted in everyday life have a geometric implication (how much fence to buy to enclose my lawn, how much fertilizer to buy to cover my lawn, how to find the shortest distance from home to work, etc.). At another level, the graphs we use to convey information have a geometric interpretation. In fact, geometric skills are much under-utilized in practical approaches to problem solving.

So, all the basic quantitative ideas can involve data. I would even suggest that we go one step further and try to teach people how to deal with data that comes from a sample. (Sampling variability, sampling distribution, sampling error for proportions and means are some key examples.) Polls and the interpretation of experimental data (especially health-related data) are so common in the media that it behooves us as educators to try to help people deal with the problem. (Did the recent report from the Physician's Health Study actually say everyone should take aspirin to reduce the risk of heart attack?) These ideas can also be conveyed without the formal structure of confidence intervals and hypothesis tests.

A related goal should be in the area of "Helping people to be life-long learners." I'm not sure how one goes about doing this, but part of it must involve inciting curiosity about numerical phenomena. I think people tend to be curious about numerical things (what are my chances of winning the lottery?) but think the solutions are somehow mysterious and beyond them. Thus, we need to provide simple ideas for handling numbers that do not involve messy algorithms and can be seen as widely applicable building blocks to number sense. (A jackknife, not a surgeons scalpel?)

I have not addressed, except indirectly, the issue of appreciation of mathematics. I believe this is important so that parents, legislators, and others can evaluate the importance of mathematics education. Although I have not thought through this idea real carefully, I think appreciation can be taught through data as well. One way to extend the elementary notions of data analysis mentioned above is through the building of models. We can see that aspirin reduces the risk of heart attack in some people, but a more comprehensive view of how the risk is affected by age, weight, race, and a myriad of other variables can only be obtained through a statistical model. We can see data on the reliability of a component of a space shuttle but the reliability of the entire shuttle can be estimated only through a complex mathematical model. Perhaps students can learn to appreciate the power of mathematics by understanding the simpler components well enough to envision the role of a more detailed model, even if they have no interest in constructing the model themselves.

—Richard Scheaffer, University of Florida

Multiculturalism in Mathematics

Multiculturalism in Mathematics:
Historical Truth or Political Correctness?

Allyn Jackson

PROVIDENCE, RHODE ISLAND

Introduction

Multiculturalism, cultural relativism, Eurocentrism, political correctness—these are fighting words in academia today. As conservatives clash with liberals over who controls the collegiate curriculum, the fight over multiculturalism is shaping much of educational debate today. With books like Dinesh d'Souza's *Illiberal Education* running up against a growing Afrocentrism movement in academia, the debate has been scorching. As one writer put it in an opinion column in the *Chronicle of Higher Education*, "Multiculturalism has emerged suddenly as a scholarly buzz word, and, with equal speed, has been converted into an epithet."

The term multiculturalism is used in many different ways to encompass many different aspects of teaching and scholarship. Generally, it refers to attempts to acknowledge and investigate the role of all cultures having an influence on a particular field. As it pertains to the collegiate curriculum, multiculturalism has perhaps become most closely identified with attempts to revise the literary canon of humanities courses to include authors from varied ethnic backgrounds. Charging "political correctness," conservative scholars have decried this trend, saying it stifles debate and dilutes the impact of education by including every conceivable viewpoint. Liberals shoot back, saying that traditional curricula have stifled debate for years by giving credence to only Western viewpoints and values.

What does all this have to do with mathematics? Plenty, though the connection might surprise some. Many academic mathematicians would probably have a reaction similar to that of one prominent mathematics professor at a well-known private university, when faced with the term "multiculturalism in mathematics:" "What?! If there's anything that's neutral and independent of culture, that has nothing to do with ethnicity or gender, it's mathematics!" Another mathematician laughed knowingly at the phrase: he had recently been asked by a gay activist group on campus for information on the work of homosexual mathematicians. In fact, it appears that the impact of multiculturalism on collegiate mathematics has for the most part been very small. But, judging from talks presented at conferences, articles appearing in journals, and books being written, it's clear that a multicultural outlook is gaining ground as a way of viewing mathematics.

Discerning the general opinion of this area is difficult, because many in the collegiate mathematical community don't know much about it. The enthusiasm of its proponents is clear, but equally clear is a strong undercurrent of suspicion that many are reluctant to voice. Some of those interviewed for this article who were the most negative about multiculturalism in mathematics would not go on record with their views. Some said that the entire area had not produced worthwhile results but were unwilling to criticize it publicly because they felt that the field was just beginning. Off the record, a few referred negatively to groups who were "just trying to show that everything started in Africa." One critic who also would not be quoted by name had this to say: "It's bad history and bad mathematics tied together with a hefty dose of racism."

Is Mathematics Culture-Free?

The various attempts to connect mathematics with culture have given rise to a wide range of ideas and approaches. Prominent among these is ethnomathematics, a scholarly endeavor, anthropological in spirit, which examines mathematical ideas as they arise in a particular historical, cultural, political, or social context. Often, ethnomathematics is tied to education and pedagogy in the belief that, in order to teach mathematics, one must understand the mathematical knowledge students bring with them from their own cultures. A related area is humanistic mathematics, which attempts to recognize and emphasize the role of intuition, discovery, emotions, and values in the development of mathematics. For the most part, humanistic mathematics has educational goals, but it also connects to ethnomathematics in its emphasis on the human aspect of mathematical ideas. Usually, multiculturalism is understood to refer to educational efforts to broaden the scope of mathematics teaching and learning to encompass mathematical ideas that arise in a range of cultures.

These ways of looking at mathematics immediately run into a big philosophical question: is mathematics culture-independent, or does it have "social fingerprints" from the environment in which it was developed? Most people would agree that the concept embodied by the statement $2 + 2 = 4$ has an inherent truth independent of any human uses to which the concept might be put. Unlike other areas of human endeavor, mathematical ideas possess a unique internal consistency. "It is interesting to talk about mathematics as if it were in the same category as art, or philosophy, or social science," points out Reuben Hersh, who is on the mathematics faculty at the University of New Mexico. "In those areas, it's very clear that different countries have different versions and they can even conflict. But mathematics does not conflict from one country to another... When there seems to be a conflict, and the people involved actually communicate, it's straightened out pretty soon. This doesn't happen in sociology and art and so on. So, I think as far as the content of mathematics is concerned... it's pretty objective, in the sense of being culture-independent. However, that isn't necessarily the case for style of presentation in writing and lecturing."

Hersh's colleague and co-author, Philip J. Davis, takes a somewhat different view. Davis, who is on the applied mathematics faculty at Brown University, wrote in the *Humanistic Mathematics Newsletter* that the commonly held Platonic view is that a statement like $2 + 2 = 4$ "is perfect in its precision and in its truth, is absolute in its objectivity, is universally interpretable, is eternally valid, and expresses something that must be true in this world and in all possible worlds." He then points out that extending this view beyond simple arithmetic truths is problematic: "the obvious fact [is] that we humans have invented or discovered mathematics, that we have installed mathematics in a variety of places both in the arrangements of our daily lives and in our attempts to understand the physical world... Opposed to the Platonic view is the view that a mathematical experience combines the external world with our interpretation of it, via the particular structure of our brains and senses, and through our interaction with one another as communicating, reasoning being organized into social groups."

The interplay between the facts of mathematics and social forces have clearly influenced the kinds of mathematical questions that get asked. Mathematical historian Judith Grabiner of Pitzer College maintains that the social influences on mathematics have been "immense." "There's been a great deal of interaction between social context and mathematics research," she notes. "That doesn't determine that $2 + 2 = 4$, but it does affect the directions of mathematics research." As an example, she points to Huygens' work on the theory of evolutes in curves. "The theory of evolutes comes from his interest in the pendulum clock, and his interest in the clock comes from his interest in the problem of longitudes of ships at sea," she explains. "If Holland had been a land-locked desert nation, I doubt that this problem would have interested him." Mathematics has even been shaped by religious practices. For example, followers of Islam must pray in the direction of Mecca. Not content with praying in a direction that was "vaguely east," the Muslims "took it more precisely, so you need some spherical trigonometry to find out the direction of Mecca... It's a nontrivial mathematical problem, in the context of the eighth and ninth century... There's an example of a religious need that gave rise to a direction of mathematical research."

As a more current example, some argue that research funding applies political pressure to the directions mathematical research takes. "It's not a coincidence that some fields are more heavily funded than others," says Claudia Henrion, a mathematician at Middlebury College who has an interest in women in mathematics and in social and political aspects of the field more generally. "There is a lot of resistance within mathematics to ask those kinds of questions. There's a really strong belief that you can separate the doing of mathematics from social and political issues." "To say that there are no social fingerprints in mathematics defines what are legitimate questions to ask about mathematics," she continues. "And it makes it okay to ask, 'Is Fermat's Last Theorem true?', but not okay to ask, 'Why were certain classes of people excluded from the world of mathematics?'"

Some go a step further and question the cultural neutrality of such underlying notions of Western mathematics as rationalism and abstraction. In "Western Mathematics: The Secret Weapon of Cultural Imperialism" (*Race and Class*, 32 (2) 1990), Alan J. Bishop of the Department of Education at Cambridge University "challenges the myth" that mathematics is "a culturally neutral phenomenon in the otherwise turbulent waters of education and imperialism." To support his argument, he looks at mathematics in trade, administration, and education in countries where the culture of a colonial power was imposed on the indigenous people. He concludes that "western mathematics presents a dehumanized, objectified, ideological world-view" that is at odds with many of the values and modes of thought of other cultures. Noting that Western mathematics is now taught in every country of the world, he remarks that "the cultural imperialism of western mathematics has yet to be fully realized and understood. Gradually, greater understanding of its impact is being acquired, but one must wonder whether its all-pervading influence is now out of control."

Broadening History

Whether mathematics is culture-independent is a question not likely to be definitively resolved. However, most would agree that the investigation and documentation of mathematical ideas arising in other cultures has been a neglected aspect in the history of the development of mathematics. Broadening the history of mathematics to include cultures outside of western Europe is a major goal of ethnomathematics. By most accounts, this term was coined by Ubiratan d'Ambrosio of the Universidade Estadual de Campinas in Brazil. According to d'Ambrosio, ethnomathematics provides a framework for examining the different ways people use mathematical thinking. "Classifying, measuring, estimating, counting, and so on, these are all techniques to understand and cope with reality, and they are found in all cultures that we have records for," he points out. "'Ethno' emphasizes that these ways of explaining and understanding are based on social and cultural roots." D'Ambrosio came to ethnomathematics through studying the history and philosophy of mathematics. But he found the prevailing views "so Eurocentric, as if the only value, the only way of thinking, comes from Europe. I wanted to see history in a broader way." His main concern, he says, "is to make ethnomathematics a scholarly, respected field with all the usual standards of scholarship." One form of recognition came recently when ethnomathematics was placed under the History and Biography section of the 1991 Mathematical Reviews Subject Classification.

Another sign of growth in the field is the formation in 1985 of the International Study Group on Ethnomathematics (ISGEm). ISGEm is mainly concerned with sharing information and fostering collaborations among its members, who are scattered all over the globe. The study group focuses on four main aspects of ethnomathematics: theoretical perspectives, research in culturally diverse environments, out-of-school applications, and curricular and classroom applications. Patrick Scott, a mathematics education professor at the University of New Mexico, is the editor of the ISGEm newsletter. The newsletter is mailed to about 200 people, but it actually has a wider circulation: because there is interest in the newsletter in dollar-poor countries, a number of representatives around the world photocopy the newsletter and send it to colleagues who would have difficulty paying for it. Among the topics covered in the newsletter are ways to use video games to illustrate mathematical topics, a report on ethnomathematical investigations of a wide variety of cultural activities in Mozambique, and an analysis of connections between ethnomathematics and the different "learning styles" of majority and minority students.

Heading ISGEm is Gloria Gilmer, President of Math-Tech, Inc., an educational research and development corporation in Milwaukee. Gilmer does ethnographic research on the mathematical activities of inner-city groups, with the goal of helping these groups make the connection to more abstract mathematical concepts. She sees ethnomathematics as a way of broadening the kinds of activities that are considered to be mathematical. "The question is, is it mathematics if it's not in the form that you ordinarily think of as mathematics?" she asks. "Or are we bent on calling certain things mathematics because of a certain form that we're imposing on the discipline?" The mathematics that most people are familiar with, she notes, is very formal and academic. "We think there is something beyond this, and we would like to call this ethnomathematics."

In contrast to Gilmer's work, some of the research in ethnomathematics is concerned with the mathematical ideas that arise in nonliterate cultures. *Ethnomathematics* (Brooks-Cole, 1991), by Marcia Ascher of the mathematics department at Ithaca College, is a good introduction to this aspect of the field. The book has been recognized for its careful scholarship and its wide range of subject matter. She writes that "Ethnomathematics, as it is being addressed here, has the goal of broadening the history of mathematics to one that has a multicultural, global perspective." To work toward this goal, Ascher focuses on several different instances in which mathematical ideas arise in the traditions of nonliterate cultures.

The mathematical constructs she discusses are remarkably sophisticated. For example, she investigates the sand tracing tradition among the Malekula of the New Hebrides. In the Malekulan mythology of death, these sand tracings play an important part in reaching the Land of the Dead, and it turns out that the tracings they do are Eulerian paths, the same graph-theoretic construction as found in the famous problem of the seven Königsberg bridges. Similar sand tracings can also be found among the Bushoong of Zaire and the Tshokwe of Central Africa. Another section of the book focuses on certain indigenous Australian groups in which kinship relations are organized using a dihedral group of order eight. Games of chance and strategy are analyzed, including what Philip Morrison, in a laudatory review of the book in *Scientific American,* calls a "kind of hyperbackgammon played with counters on an eight-pointed star."

Is Traditional History Eurocentric?

Some mathematical historians have worked on understanding the mathematics of other cultures but do not refer to their work as ethnomathematics. For example, David Pingree, a mathematical historian at Brown University, has built upon the work of Otto Neugebauer and others in ancient mathematical astronomy. Pingree agrees with the proponents of ethnomathematics that much of the traditional history has been "Eurocentric" in taking Greece as the starting point of all of mathematics. For example, he says that the development of trigonometry, commonly attributed to Western mathematics, actually has a circuitous history beginning with what's known as the "chord function," used by Ptolemy in the second century for problems in mathematical astronomy. The sine and cosine functions were developed from the chord function in India and were transmitted to the Babylonians, who in the ninth, tenth, and eleventh centuries developed ways of computing with the functions, thereby laying the foundation for Renaissance trigonometry. In fact, he says, transmission of Arabic mathematics through Muslim Spain was a major route by which mathematical ideas entered Europe.

In reading ancient mathematical texts in their original languages, such as Cuneiform and Sanskrit, Pingree says he finds many different ways of doing science and mathematics that developed in response to the needs of various cultures. "Material was transmitted over and over again from one culture to another and reworked into something new—it's a very complex history," he says. "What we're dealing with is not some kind of simple contrast between Babylonian and Greek mathematics... While [various cultures] influenced each other, the mathematical learning that each received from the other was reshaped to fit the needs of the society that received it." It is important to understand the mathematics in its cultural context, he says. "One should simply sit down and try to figure out what, say, Indian mathematicians were doing in terms of what they were trying to do and what they achieved," he says. "And this is remarkable enough, one doesn't have to compare it to Western mathematics."

However, Pingree says, some attempts to investigate the mathematics of other cultures have, paradoxically, held Western mathematics as the ultimate yardstick against which other contributions are measured. For example, he points out that the Indian mathematician Mādhava discovered the infinite series for the sine, cosine, and tangent functions around

1400, centuries before Gregory, the European mathematician generally credited with the discovery. Mādhava's demonstration was not completely rigorous by today's mathematical standards, but it was nonetheless quite correct. (In fact, Pingree says that some of his colleagues in the Brown mathematics department use Mādhava's demonstration in their classes because its geometric approach is more intuitive.) Pingree says that some Indian scholars have renamed the series as the Mādhava-Gregory series. "They're simply using the name, I don't begrudge them that," he says. "But in a sense this implies that Mādhava's understanding of the series was the same as Gregory's, which is not true. And there are other Indian scholars who will go whole hog and talk about how Mādhava discovered the calculus. This is simply false."

This kind of viewpoint, says Pingree, implies that scientific or mathematical developments have value and interest only if they can be related to something in Western culture. "This is one of the terrible things that I think is prevalent in many countries where the cultures have been overwhelmed by the impact of the West," he notes. "And some of the people who are affected by this respond by trying to claim their culture was within it the terms of Western science." In addition, he notes that such scholars often emphasize only those ideas that supported Western science, while ignoring later results that went in other directions.

As a result, Pingree is quite skeptical of attempts to multiculturalize mathematics history, which he believes are often motivated by political aims. For example, he serves on a committee of the Third World Academy of Sciences that gives a prize for an article on the history of science in the Third World. He says submissions by Third World writers have never received the prize because their work is generally written with a "nationalistic" viewpoint. These kinds of writings, he says, try "to prove the superiority of their culture by claiming that it pre-dates Western science. If that is their basis for their feeling good about their culture, that's a very sad comment on the total destruction of that culture."

Connections to Education

In the area of mathematics education, multiculturalism has had an influence primarily at the school level; the impact on collegiate-level teaching and curricula has generally been very small. For one thing, from calculus onward, the courses all concern mathematics as it developed in the West. However, some faculty are incorporating ideas of multiculturalism into certain courses, primarily those on the history of mathematics.

For example, Claudia Henrion of Middlebury College teaches a history of mathematics course with an emphasis on connections to social and cultural questions. The first part of the course looks at the history of "mathematical people"—who went into the field of mathematics and why, what their lives were like, and how their social environment influenced them. The second part of the course focuses on the relationship between mathematics and culture by looking at African, Egyptian, Babylonian, Greek, Chinese, Indian, and Muslim mathematics. Henrion says that looking at the mathematics that developed in different cultural contexts is "useful in seeing how mathematics is not the same in different cultures, and in trying to understand the relationship between the values of a society and the kinds of mathematics they choose to develop." The third part of the course brings together ideas developed in the first two parts to examine the history of geometry.

Out of this course grew an idea for a regional conference in New England, to be held in June 1993, on mathematics as a human activity. Henrion says that the purpose of the conference is to bring together mathematics professors to "talk about how to add the 'hidden' dimension of mathematics to their course in different ways and to understand the relationship between people and mathematics, and culture and mathematics." In addition, the conference will look at parallels between mathematics and the humanities and the implications these kinds of ideas hold for mathematical pedagogy.

At Ithaca College, Marcia Ascher has developed a course for upper freshmen to seniors in any major; she says she has had students who range from typical "math-avoiders" to seniors majoring in mathematics and physics. Her book *Ethnomathematics* contains many of the topics and ideas she covers in her course. For example, one topic is strip patterns, which are repeating decorative patterns that run over a two-dimensional strip of fixed width. The students first pick a basic repeating unit and generate their own strip patterns, and then they work on a project to find and analyze strip patterns in other cultures. Some interesting group theory comes into play here, because it turns out that all strip patterns can be analyzed using seven symmetry groups. Ascher says that the students gain an aesthetic appreciation in delving into both the beauty of the patterns and their internal logic. "We have a mathematical abstraction of symmetry groups from strip patterns," she notes. The project allows the students to see "these diverse expressions of mathematical abstraction in human settings."

One indication that these kinds of ideas are gaining ground in the mathematical community is the development of the Humanistic Mathematics Network, run by Alvin White of the mathematics department at Harvey Mudd College. The Network's main activities are organizing of sessions at national meetings and publication of a newsletter, which began in 1987 with a grant from Exxon Education Foundation. The Network has served as an umbrella for all kinds of discussions—from ethnomathematics, to the philosophy of mathematics, to ethical questions, to women in mathematics. Judging from the articles in the newsletter, there is a great deal of interest in the mathematical community in finding ways to emphasize the human aspects of mathematics as a means of connecting students to the mathematics they're learning. Some of the activity of the Network has centered on multiculturalism and ethnomathematics as a way to do this.

Although multiculturalism is often seen as a way to bring in students from groups traditionally under-represented in mathematics, it is interesting to note that many of the nationally-known programs to promote the achievement of minority students in mathematics have no explicit multicultural aspects. The Alliance to Involve Minorities in Mathematics at Mathematical Sciences Education Board of the National Research Council, headed by Beverly Anderson, professor of mathematics at the University of the District of Columbia; the Quality Education for Minorities Network, headed by Shirley McBay, a mathematician and former dean of student affairs at MIT; the Dana Center for Science and Mathematics Education, headed by Uri Treisman, professor of mathematics at the University of Texas at Austin; the TexPREP program, headed by Manuel Berriozábal, professor of mathematics at the University of Texas at San Antonio; the SUMMA program at the Mathematical Association of America, headed by William Hawkins, professor of mathematics at the University of the District of Columbia—all of these programs emphasize the importance of role models, support mechanisms, intellectual challenges, high-quality instruction, and hard work, but do not specifically embrace multiculturalism as a major aspect of what they do.

Berriozábal probably reflects the thinking of many of these programs when he says that mathematics is an intellectual challenge that "requires persistence, interest, and hard work." Because he teaches mostly Hispanic students, he says it is "fair, right, and appropriate" to bring in references to the achievements of such indigenous peoples as the Aztecs and the Mayans, "to give students an appreciation of the mathematics of other cultures." In European civilizations, technology did grow faster than in other places, he points out. But "today, in the civilizations we're living in, Hispanic and African-American youngsters are part of our civilization and there is no reason to think they can't be contributors." It's important to be inclusive of various groups, "but, as far as becoming a mathematician, there is nothing that substitutes for hard work and achievement."

An Integrated Approach

One of the few examples where culture, history, and heritage are woven into mathematics teaching at the collegiate level is found at Clark Atlanta University, a Historically Black Institution in Atlanta, Georgia. The driving force behind this effort is Abdulalim Shabazz, chair of the mathematics department at Clark Atlanta. "We live in a society that believes that heredity plays a part in who is going to be on the top and who is going to be on the bottom," Shabazz notes. "I say that's hogwash, that's nonsense." He says the common perception is that most students cannot excel in advanced levels of mathematics and that, in particular, "African-Americans, Hispanic-Americans, and Native-Americans are not expected to excel in mathematics. Now those perceptions, I let my students and my colleagues know, are false... Everyone who has the desire and who has the spirit of work, hard work, can reach high levels of achievement in the mathematical sciences."

Shabazz tries to create in the department an environment of cooperative learning, discussion, and debate. From what he describes, there seems to be a great deal of enthusiasm for mathematics among the students. He says that a visitor coming to the mathematics building would find that "we don't have the kinds of space, we don't have the materials that others have," he says. "But the students are studying, doing mathematics all around the hallways... We don't have lounges, so we have transformed the hallways into tutorials and places of discussion for mathematics and mathematical debate."

Telling students about the contributions their ancestors made to mathematics is an important component of Shabazz's approach. The objective is to "let them know that in their past, in their history, in their heritage, they have achieved greatness in the mathematical sciences," he declares. "In doing this, I simply tell them the truth. The origins of the mathematical sciences are to be found in Africa." He says he does not espouse an Afrocentric position, but rather a "truth-centric" position. "For example, in the Eurocentric teaching, they speak about mathematics beginning with the Greeks, which is not true," he says. "The Greeks were educated in Africa. And it was from Africa that Thales and Pythagoras and Oenipides and Democritus and Euxodus and Plato—all of these people came to get knowledge, and then of course it spread into the European continent. Euclid, who was an African, born in Africa, never left Africa, is always painted as a European... And Hypatia, an African woman [mathematician]... I've seen some pictures of her, and they have her looking like a European. In some cases, she looks like a Hollywood starlet! She was born in Africa, never left Africa."

Shabazz says this approach makes a big difference to his students, because the prevailing messages are that Black people played no part whatsoever in the development of mathematics. "It has to do with the question of self-esteem," he says. "In teaching students, you have to keep their self-esteem intact... I tell them, remember, your forefathers were the ones that built the pyramids, who put the Sphinx out there on the desert, in Egypt. Not only there, but you have pyramids in this hemisphere, the Americas... And then show them, by the existence of the huge African heads that have been found throughout various places in Central America, that Africans were traveling back and forth across the Atlantic long before Columbus ever dreamed of sailing the ocean blue."

This is Shabazz's second time teaching at Clark Atlanta University—he taught there from 1957 to 1963, returning in 1986 after nearly two decades as an Islamic preacher and four years as a mathematics professor at the University of Mecca. During his first period at Clark Atlanta, he produced impressive results—the department awarded 109 master's degrees in mathematics to Black students. Over a third of them went on to get doctorates in mathematics or mathematics education, and many of them in turn produced other Black Ph.D.s in mathematics. He reports that there have been estimates that approximately one-half of the estimated 200 Black mathematicians in the U.S. were either students of his or students of his students.

Why Multiculturalism?

Why introduce students to the mathematics of other cultures? Claudia Zaslavsky, a mathematics educator whose 1973 book *Africa Counts* has become a classic of ethnomathematics, provides a cogent summary of the various reasons in an article in the ISGEm newsletter: "[Students] learn that mathematical practices arose out of the real needs and desires of all societies. Mathematics comes alive when children study the measurement and numeration systems, the patterns in art and architecture, the games of skill and games of chance of various cultures. Students have the opportunity to learn about mathematical contributions of women and of Third World societies, a generally neglected area of mathematics. They can take pride in their own heritage, and at the same time become familiar with and learn to respect the cultures of other societies."

Most who work in ethnomathematics see these educational dimensions as important. In writing *Ethnomathematics*, Ascher says she was not motivated primarily by educational concerns, but she believes her work can function as an educational tool. "My concern really is to enlarge the vision of mathematics for mathematicians and for students they teach so that it does have a global perspective," she notes. "When you look in math history books... non-Western peoples, if they appear, usually appear in an aside or as basically trivial... There is no question that if the only place you see yourself in mathematics is in some quaint story about how Africans can't count to two or something, it's not a very friendly greeting."

"It's clear that some of the problems in schools are not a question of method or curriculum, but occur because of cultural distance between what we want to teach and the student," notes d'Ambrosio. "Ethnomathematics is trying to bridge this gap." He says that he has given talks in many different countries, and his message is well-received because many people are recognizing that the current mathematics educational practices are producing disappointing results. "The students are smart, but why do they react so badly to mathematics?" he asks. "It's because they find there is no cultural appeal to it. They find it remote from their own concerns and their cultural environment."

Understanding and connecting to this environment is an important aspect in teaching mathematics, says Gilmer, noting that teachers must provide a context for mathematics that engages students. "The context engages them, and when they are engaged, they think," she declares. "Some problems don't engage them, they're not interested... We have a lot of sterile problems, like 'add these monomials.' Well, [the students] get all kinds of weird answers on that because it doesn't mean anything to them." Because of the range and breadth of mathematics, one can always find problems that will engage students, she says. "To me, mathematics is extraordinarily flexible. The context of the problem can always be shifted. Mathematics interfaces with ordinary life in so many ways, we don't have to be stilted in formulating problems for students."

Often, introducing multicultural elements is intended to produce a "cultural reaffirmation" among the students. Patrick Scott of the University of New Mexico worked in Guatemala with Mayan children in a program intended to bring elements from the indigenous culture into teaching mathematics. "One purpose of using Mayan mathematics with Mayan children isn't so much the content of the mathematics as it is a cultural reaffirmation to make them feel good about themselves and their ability to learn mathematics and their own culture's contribution to mathematics," he explains. "Often, with the Eurocentric emphasis that there has been in mathematics, students get the impression that mathematics is something of the white man, that only the white man developed it and only the white man can use it... And then, I think that, for students from any culture, it can be both entertaining and edifying to see what mathematics has been developed in other cultures."

Judith Grabiner of Pitzer College agrees with this general viewpoint. For example, she says that she told a precalculus class that Medieval Islam made important contributions to the development of algebra and that the word algorithm comes from the name of a 9th-century Islamic mathematician, al-Khwarizmi. At the end of the course, one Black student signed his course evaluation and thanked Grabiner for refuting the myth that all mathematics was created by white Europeans. Such details "tell the students, this is yours too," she notes. "It's not that they can't learn mathematics in the traditional way. But you can make the classroom a friendlier place... It gives the students a sense of shared heritage."

However, Grabiner is quick to point out the importance of providing accurate information. "There is a danger in ethnomathematics of getting things wrong and saying that people have done things they haven't," she explains. For example, "it is important to know that the Babylonians knew of and used the fact that the sum of the squares of two legs of a right triangle is equal to the square of its hypotenuse." She says that some have claimed it's wrong to label this fact as the Pythagorean Theorem. But there is a good reason for the name, she declares: "The Pythagoreans proved it logically." "It's important to be accurate," she says. "The wonderful things people have created are already great, and it doesn't do anybody any good to exaggerate."

Some multiculturalization efforts have been criticized on exactly these grounds. For example, one document that has come under fire for inaccuracies is the "African-American Baseline Essays," compiled for the public schools in Portland, Oregon; these essays have also been used in other school systems, including Atlanta, Detroit, and Washington, DC. There are six essays on art, language arts, music, social science, science, and mathematics. The essays are intended to provide information about the history, culture, and contributions of Africans and African-Americans and are to be used by teachers and other educational staff as a reference and resource in developing curricula and classroom materials.

The science essay has been heavily criticized for inaccuracies and for containing a great deal of material that ordinarily wouldn't be considered science at all, such as descriptions of Egyptian religious beliefs. The essay claims that Egyptians used parapsychology as a scientific method and constructed gliders that they flew in for travel, expeditions, and recreation—the *Guinness Book of World Records* is cited to support the latter claim. In addition, the essay claims that Egyptian physicians treated "trans-material" aspects of illnesses by transferring "[energy] which they received from the Pharaoh, who in turn received his [energy] from the sun, to the patient." The mathematics essay does not contain these kinds of obvious problems and does not seem to have been the subject of criticism. However, it does contain some statements that many historians disagree with, such as that Euclid was born and lived his entire life in Africa and that there is "no logical reason" for the appellation "Pythagorean Theorem."

For all the efforts to connect mathematics to culture, a paradox remains: mathematics is often found to be the academic area in which students from non-Western cultures excel, precisely because it requires so little understanding of the culture in which it is taught. Donald Crowe, a mathematician at the University of Wisconsin at Madison, has done research applying the tools of crystallography to study symmetries in patterns arising in archeological and anthropological situations, so he has some understanding of the mathematics that arises in different cultures. In the 1960s he taught at a university in Nigeria where he says "we were basically imposing an English curriculum on this Nigerian situation." The students had trouble, for example, with botany, since the textbooks discussed English plant life. But in mathematics, he said, the students did just fine. "If you're going to read Jane Austen, you have to know about the culture," he remarks. "But if you're going to prove theorems in group theory, you don't have to know anything about the culture. You have to know the axioms of the proof, and any intelligent person can learn those. And there was no question that these students were intelligent."

Despite this example, Crowe says he believes that introducing elements of the mathematics of other cultures can be valuable in stimulating students' interest in mathematics. Others are more skeptical. For example, Diane Ravitch, who is now Assistant Secretary for the Office of Educational Research and Improvement in the U.S. Department of Education, has been a harsh critic of multiculturalism in all academic areas. In her article "Multiculturalism: E Pluribus Plures" (*Key Reporter*, Autumn 1990), she argues that injecting multiculturalism into mathematics education would be disastrous at a time when educators are struggling to revise the mathematics curriculum. "If, as seems likely, ancient mathematics is taught mainly to minority children, the gap between them and middle-class white children is apt to grow," she writes. "It is worth noting that children in Korea, who score highest in mathematics on international assessments, do not study ancient Korean mathematics... We might reflect, too, on how little social prestige has been accorded in this country to immigrants from Greece and Italy, even though the achievements of their ancestors were at the heart of the classical curriculum."

George E. Andrews, professor of mathematics at Pennsylvania State University, is also unconvinced by claims that multiculturalism will help students of diverse ethnic backgrounds excel in mathematics. He believes that the "replacement of human beings by political stereotypes" underlies much of multiculturalism. "There is very little humanity in it," he asserts. "There is a view that there are the 'ins' and the 'outs,' and the important thing to do is to make sure that the 'ins' pay for their past sins, and the 'outs' are 'empowered.'" He doesn't see much to be gained in showing mathematics students the contributions that their ethnic group made to mathematics. "[This] is basically saying to the kids, you're smart because somebody who has been long dead and who is related to you was smart," he says. For example, "the suggestion that I would believe that because Newton and I have the same sex organs, I am somehow ennobled, is one of the most ridiculous ideas that I can possibly imagine. And perpetrating nonsense like this on innocent children seems to me just pathetic." He says this approach also leads to a paradoxical situation: "If, unfortunately, in your ethnic group, we're not able to find anybody back there who did anything, then you're stuck."

Andrews is also suspicious of multiculturalists' aims of promoting self-esteem among students. He points to a report in *Time* magazine about an international comparison of mathematical ability, in which Korean 13-year-olds placed first and the United States last. On a "self-esteem" question, which asked the students to say "yes" or "no" to the statement, "I am good at mathematics," the American students placed first, with 68% saying "yes," while the Korean students placed last, with just 23% saying "yes." "What the proponents of multicultural education believe is that if you promote self-esteem, you somehow promote education," he declares. In this study, "the people who really are educated come in dead last on the self-esteem question, whereas we, who have been preaching self-esteem from kindergarten through grade twelve—and now hope to perpetrate it on the universities— have absolutely mountains of self-esteem but we don't know how to do mathematics."

The sharp differences in opinion about multiculturalism aren't softened by any hard evidence—it would be impossible to devise a conclusive test of whether students learn better when culture is taken into account. Still, the arguments put forth in support of multiculturalism are persuasive to many, and, in some cases, there seem to be indications that multiculturalism can be a way to interest more students in mathematics. And, amid national dissatisfaction with how little mathematics young people are learning, the goal of bringing mathematical understanding to a wider range of students has a good deal of political force to it.

Will multiculturalism eventually have an impact on collegiate mathematics? Before that happens, mathematics faculty would have to drastically change their thinking. "I personally find ethnomathematics interesting and potentially important. But I think most mathematicians find the whole thing uninteresting and annoying," says Reuben Hersh of the University of New Mexico. "They don't want to hear about it. The idea that they would have to know about mathematics in Africa before Euclid and explain it seems very alien and unimportant. But I suppose that if it is politically important, they'll have to do it—someone will tell them they must." Wouldn't that raise a controversy? "Well, there already is a controversy. The only thing is, the controversy... has not reached a level where people have to pay attention to it."

Appendix: Reading List

The following list of readings related to multiculturalism, ethnomathematics, and the contributions to mathematics of non-European cultures was prepared by Dr. Florence Fasanelli of the Mathematical Association of America.

Ascher, Marcia. *Ethnomathematics: A Multicultural View of Mathematical Ideas.* Pacific Grove, CA: Brooks/Cole, 1991.

> An excellent detailed consideration of the mathematical ideas of people in traditional cultures. Each chapter in this 200-page volume contains notes with extensive annotated references.

Bedine, Silvio A. *The Life of Benjamin Banneker.* Rancho Cordova, CA: Landmark Enterprises, 1972.

> The definitive biography of the first Black man of science in the USA.

Berggren, J.L. *Episodes in the Mathematics of Medieval Islam.* New York: Springer-Verlag, 1986.

> Very readable survey of medieval Arabic mathematics.

Bos, H.J.M. and Mehrtens, H. "The interactions of mathematics and society in history: Some exploratory remarks." *Historia Mathematica,* 4 (1977) 7–30.

> Arguments for studying the social history of mathematics.

Closs, Michael P. (Ed.). *Native American Mathematics.* Austin, TX: University of Texas Press, 1986.

> Recent research on early mathematics in the Americas.

D'Ambrosio, Ubiritan. "Ethnomathematics and its place in the history and pedagogy of mathematics." *For The Learning of Mathematics,* 5 (1985) 44–48.

D'Ambrosio, Ubiritan. "Ethnomathematics: A research program in the history of ideas and of cognition." *ISGEm Newsletter,* 4 (1988) 5–8.

D'Amborsio, Ubiritan. *Socio-Cultural Bases for Mathematics Education.* Brazil: UNI-CAMP, 1985.

> D'Ambrosio, the father of ethnomathematics, has written numerous articles and books on its different aspects, some of which are in English. Each of the references listed here contain bibliographies of some of his other writings.

Grabiner, Judith. "The centrality of mathematics in the history of Western thought." *Mathematics Magazine,* 61 (1988) 220–230.

> An excellent presentation of the relationship between Western mathematics and Western thought.

Joseph, George. *Crest of the Peacock: The Non-European Roots of Mathematics.* New York: St. Martin's Press, 1991.

> The best elementary study of Chinese, Islamic, and Indian mathematics available.

Lumpkin, Beatrice. *Senefer and Hatshepsut.* Chicago, IL: DuSable Museum Press, 1983.

> A tale of how to teach young children by using ancient Egyptian unit fractions.

Morrison, Philip. "Book Reviews." *Scientific American*, 8 (1991) 108.

 A review of Ascher's *Ethnomathematics* stating some of its principles with noteworthy clarity.

Swetz, Frank J. and Kao, T.I. *Was Pythagoras Chinese?* Philadelphia, PA: Pennsylvania University Press; and Washington, DC: National Council of Teachers of Mathematics, 1977.

Zaslavsky, Claudia. *Africa Counts: Number and Pattern in African Culture.* Boston, MA: Prindle, Weber, and Schmidt, 1973.

 The hardcover edition is out of print but fortunately this classic is available in paperback through Lawrence Hill Books, Brooklyn, NY 11238.

NEWSLETTERS

The newsletters listed below regularly contain articles on multiculturalism as well as bibliographies, book reviews, articles, recent research, and announcements of national and international meetings.

Humanistic Mathematics Network Newsletter. (1987–) Alvin White (Ed.). Harvey Mudd College, Claremont, CA, 91711.

ISGEm Newsletter: International Study Group on Ethnomathematics. (1985–) Patrick Scott (Ed.). College of Education, University of New Mexico, Albuquerque, NM, 78131.

International Study Group on the Relations Between the History and Pedagogy of Mathematics Newsletter. (1973–) Victor Katz (Ed.). Department of Mathematics, University of the District of Columbia, 4200 Connecticut Avenue NW, Washington, DC 20008.

AMUCHMA Newsletter: The African Mathematical Union Commission on the History of African Mathematics. Paulus Gerdes (Ed.). Faculty of Mathematics, Higher Pedagogical Institute, C.P. 3276, Maputo, Mozambique.

Assesment of Undergraduate Mathematics

Assessment of Undergraduate Mathematics

Bernard L. Madison

UNIVERSITY OF ARKANSAS

> We are between the proverbial rock (externally designed assessment) and the hard place (internally designed assessment). We must either turn over the assessment business to those who don't teach mathematics or spend a great deal of time and effort to discover how to do something about assessment ourselves. My hope lies in the community of mathematicians, but there will be no three-line proofs that solve this problem.
>
> —*Hank Frandsen, University of Tennessee, 1991*

This paper is a result of an electronic mail discussion conducted in April and May, 1991. The fourteen participants in the electronic mail focus group consisted of twelve mathematics or mathematics education faculty members and two persons from outside mathematics. Four of the mathematics faculty members now hold university administrative positions outside the mathematics department. Experience in assessment ranged from near zero to several years of university-wide coordination of assessment activities. The institutions of the participants were at various stages of assessment, ranging from having established full-scale comprehensive programs to only thinking about and discussing assessment generally. Some felt that their institutions would not institute assessment programs unless they were mandated, and those who had comprehensive programs were operating under governing board mandates. The institutions represented included small private colleges, regional comprehensive public universities, and large public research institutions.

This paper is guided more by the content of the electronic mail discussion than by the current literature, theory, and practice in student outcomes assessment. The references at the end of the paper will lead the reader to other discussions.

Evolution of Assessment

Traditional assessment in U.S. higher education splits into two modes, more or less historically. In early U.S. education, comprehensive end-of-program examinations were common, essentially the rule. External examiners of students, following the European tradition, were used by most institutions. As both geographic separation and numbers of students increased, external examiners and comprehensive examinations became less common. Individual course examinations and grades became the common method of assessment. Programs were evaluated on reputation, curriculum, and available resources, under the assumption that these had a high correlation with student learning.

As school enrollments, particularly collegiate enrollments, increased in the decades following World War II, the diversity of student preparation and post-secondary programs created a need for assessment for minimum competencies. Standardized tests, mostly with multiple-choice answers, emerged as the tools for measuring lower order skills for purposes of college admissions, course placements, or diagnostic teaching decisions. This situation was probably more a consequence of the available tools than of the effectiveness of the assessment using these tools. The focus was clearly on individual student assessment for making decisions about an individual student.

Program evaluation and faculty evaluation were not central issues in minimum competencies assessments. Program evaluation continued using reputation, curriculum, available resources, and scattered information about successes in student preparation. The competencies of entering students, however, became an ingredient in program evaluation, but "value added" measures were not common.

> The public has the right to know what it is getting for its expenditure of tax resources; the public has a right to know and understand the quality of undergraduate education that young people receive from publicly funded colleges and universities. They have a right to know that their resources are being wisely invested and committed.
>
> —*John Ashcroft [1], Governor of Missouri, 1986*

The second wave of assessment, coming mostly within the last decade, was principally based on using assessment of student learning as a means of program and faculty evaluation. In some cases, allocation of resources is affected by the results of assessments of student learning. Presumably, more student learning implies more resources in these cases. The major impetus for this type of assessment was accountability of institutions and programs. Much of the assessment has been mandated by governing units, from state governments down to university administrations. The Education Commission of the States reports that more than three-fourths of the states have a student assessment effort planned or in place. The American Council on Education reports that a majority of colleges and universities is in some stage of developing student assessment programs.

The future of assessment in U.S. higher education seems sure to bring new and different methods. Hardly anyone is content with our current ability to measure success in achieving educational goals. This is especially true in the liberal arts, where, as Grant Wiggins [9] has said, assessment needs to be built upon the distinction between wisdom and knowledge. (Wiggins' paper is reprinted as Appendix A to this report.) Mathematics faculty in higher education have little experience in setting the educational goals that are necessary for the current assessment movement. Our curricula are designed from courses, and educational goals, frequently unarticulated, are derived from courses and not the other way around. Either we have to come to grips with adapting our circumstances to the current assessment movement or we have to design new assessment tools that will be more adaptable to what we believe to be our (as of yet unarticulated) educational goals in undergraduate mathematics.

The Current Assessment Movement

A simple and favorable description of the new assessment mode is the attempt by an institution or a program to answer periodically three questions:

- What are we trying to do?
- How are we doing?
- What should we change?

One focus group participant expressed these questions as follows: "I describe [our assessment] program by asking three questions: (1) What does your department intend to do for its students? (2) What evidence is there that you do it well? (3) What could be done by you (and the Dean) to help you do it better?" Of course, that is simpler sounding than it really is. The questions are enormously difficult to answer and, in the general setting, encompass all of educational theory and practice. However, for an individual program or collegiate major, partial answers are tractable and far better than no answers at all.

The nature of the evidence presented to answer the second question is what separates student outcomes assessment from the traditional program reviews. Traditional program reviews have focused on questions of how the local curriculum compares to a national or common curriculum, or whether the library resources are adequate. Student learning outcomes is a new ingredient in program reviews, although data on competencies of entering students have been considered in the past. Some now see assessment (testing) of student learning as an integral part of program evaluation. One participant wrote, "I bind student assessment with program evaluation. I'm not much of a believer in regularly giving students tests and then publishing the results of those tests without giving interpretations and conclusions based on those tests."

> Where assessment measures have a regulatory, budgetary, or even public relations purpose, they are likely to develop ...disproportionate influence.
> —*Ernst Benjamin [4], AAUP, 1990*

Student outcomes assessment in higher education is conducted at several levels: basic skills, core curriculum, end-of-program, and in postgraduate workplaces. Assessment of basic skills is an extension of activity over the past two or three decades and is concerned with determining minimum competency for entry or as a benchmark for value-added testing. Assessment of learning in the core curriculum, or in general education, is frequently at the end of the first two college years and is sometimes a hurdle for entry into advanced courses. Often when this is a hurdle, the tests are labeled as "rising junior" tests. End-of-program assessment concerns both a comprehensive assessment of undergraduate learning and an assessment of learning in a major program—that is, of study in depth. The relationship between study in depth and learning in other parts of the undergraduate program are also subject to assessment at this stage. Alumni follow-up through surveys aims at determining how undergraduate learning has enabled the graduate to succeed in the workplace. A comprehensive assessment program usually includes assessment of all four types, some attitude analysis, and integration of the results.

Assessment in Mathematics

Assessment of learning in undergraduate mathematics involves all four assessment areas: basic skills, core curriculum, major, and workplace. Assessment of the undergraduate mathematics major is of primary interest to most mathematics faculty members, to members of our focus group, and to the MAA. Nevertheless, separation of assessment of the major from other types of assessment is difficult, and many of the philosophical and operational issues are the same across all types of assessment of undergraduate student learning.

Assessment of learning in the mathematics major has been generally equated with assessment in the individual courses that make up the major. Some departments have had comprehensive examinations, capstone courses, or senior projects, but most departments do not attempt any comprehensive assessment of learning in the major. The Graduate Record Examination (GRE) has provided the only significant standardized assessment tool of learning in the mathematics major, and it is focused on preparation for graduate study. Within the past five years, the Educational Testing Service (ETS) has developed a Major Field Achievement Test in mathematics. Use of this test is not widespread and many faculty members are skeptical about the independent value of a score on such a test. Little evidence of use was discovered during the focus group discussion. Said one participant, "We are not enamored of such standardized tests." Another added, "Everything I read says that these do not work. Each of our institutions is sufficiently different that there are problems with the packages."

The student outcomes assessment movement did not originate in mathematics departments, and among mathematicians there is a mixture of skepticism and optimism about the effort. Most are involved because of external mandates. In the words of one participant, "In all probability, we would not be in the assessment business had the state not mandated it." But some see the need and look forward to progress. "The purposes and motivation for assessment of undergraduate education and of the mathematics major ought to be to see whether we are making good use of the time and talents of our students and faculty to provide an education as defined in Grant Wiggins' paper [9]."

The Main Players

The players in assessment in undergraduate mathematics are students, faculty members, college and university administrators, institutional governing boards, state coordinating boards, state legislatures, and the public. Already, most states have mandated some type of assessment in higher education, and the others are likely to follow suit soon. By and large, mathematics faculty members are skeptical of this movement and see little or no benefits resulting. Students are largely unaware of the external movement and less than highly motivated to participate with best efforts. As of now, the main players seem to be university or college administrators responding to mandates from state governments. "The main players in assessment on my campus right now are administrators above the college level," said one participant. "However, the main players should be the faculty." Most agree with that sentiment and further believe that unless faculty members get involved at the planning level, there will be trouble down the road and little benefit will accrue.

The roles of various administrators differ from program to program. In one state institution where end-of-program assessment is mandated, the dean is "...in charge of making sure that something grows out of the various forms of assessments that we conduct. In some cases that has meant new positions for a department, in others it has meant summer funds for renovating the curriculum, in others it has meant partial support for a department's attempts to improve the advising of majors." In other cases, it has meant saying "no" to requests that ran counter to indications of what needed to be done in the light of assessment.

Recognizing differences in institutions and allowing for flexibility is a key to success of mandated assessment. As one participant reported, "Under the benevolent eye of the Higher Education Coordinating Board, the public institutions of higher education cooperate in developing assessment systems, but do not march to exactly the same drummer. Another commented, "Faculty attitude toward assessment here is not as negative as one might think.

The state has paid for our assessment efforts and has allowed us to design a very flexible program that lets us do many of the things we want to do anyway."

There are indications that student involvement is a problem, especially when assessment of students' learning has no direct consequences on their receiving degrees. "One of the most difficult issues that we have faced ... is getting our students to participate, and to give the assessment tests, portfolios, etc., their best shot." When tests are not a graduation requirement, then motivation is an issue. Some colleges require undergraduates to achieve a certain score on the GRE in order to graduate with a mathematics major, and some report requiring comprehensive examinations, capstone courses, or senior projects.

Establishing Goals for Assessment

> If our testing encourages smug or thoughtless mastery—and it does—we undermine the liberal arts.
>
> *—Grant Wiggins, 1990*

There is general agreement that the most difficult step in establishing a program of assessment is determining educational goals. Repeatedly, throughout the focus group discussion, the MAA was urged to publish samples of goals statements. There is no need to identify departments; in fact, the feeling is that identification tends to canonize a few departments and discourage beginners. Considerable discussion within the mathematics community over the past five years has at least laid the groundwork for formulating these goals statements. The question to answer is "What do we want our students to learn?"

The report [2] of the joint MAA and Association of American Colleges (AAC) project on study in depth addressed this issue as follows:

> Many would argue that goals for study in depth can be effective only if supported by a plan for assessment that persuasively relates the work on which students are graded to the objectives of their education. Assessment in courses and of the major as a whole should be aligned with appropriate objectives, not just with the technical details of solving equations or doing proofs. Many specific objectives can flow from the broad goals of study in depth, including solving open-ended problems; communicating mathematics effectively; close reading of technically-based material; productive techniques for contributing to group efforts; recognizing and expressing mathematical ideas embedded in other contexts. Open-ended goals require open-ended assessment mechanisms; although difficult to use and interpret, such devices yield valuable insight into how students think.

> Relatively few mathematics departments now require a formal summative evaluation of each student's major. The few that do often use the Graduate Record Examination (or an undergraduate counterpart) as an objective test, together with a local requirement for a paper, project, or presentation on some special topic. Some institutions, occasionally pressured by mandates from on high, are developing innovative means of assessment based on portfolios, outside examiner, or undergraduate research projects. Here's one example that blends a capstone course with a senior evaluation:

> > The Senior Evaluation has two major components to be completed during the fall and spring semesters of the senior year. During the fall semester the students are required to read twelve carefully selected articles and to write summaries of ten of them. (Faculty-written summaries of two articles are provided as examples.) This work comprises half the grade on the senior evaluation. During the fall semester each student chooses one article as a topic for presentation at a seminar. During the spring semester the department arranges a seminar whose initial talks are presented by members of the department as samples for

the students. At subsequent meetings, the students present their talks. Participation in the seminar comprises the other half of the grade for the Senior Evaluation.

Because of the considerable variety of goals of an undergraduate mathematics major, it is widely acknowledged that ordinary paper-and-pencil tests cannot by themselves constitute a valid assessment of the major. Although some important skills and knowledge can be measured by such tests, other objectives (e.g., oral and written communication; contributions to team work) require other methods. Some departments are beginning to explore portfolio systems in which a student submits samples of a variety of work to represent just what he or she is capable of. A portfolio system allows students the chance to put forth their best work, rather than judging them primarily on areas of weakness.

The recommendations from the National Council of Teachers of Mathematics for evaluation and assessment of school mathematics convey much wisdom that is applicable to college mathematics. Assessment must be aligned with goals of instruction. If one wants to promote higher order thinking and habits of mind suitable for effective problem solving, then these are the things that should be tested. Moreover, assessment should be an integral part of the process of instruction: it should arise in large measure out of learning environments in which the instructor can observe how students think as well as whether they can find right answers. *Assessment of undergraduate majors should be aligned with broad goals of the major: tests should stress what is most important, not just what is easiest to test.*

Obviously, one set of goals will not suffice for all programs, even for all undergraduate mathematics majors. Many have different tracks and emphases. Furthermore, mathematics faculty members are unaccustomed to viewing the undergraduate curriculum from a set of goals. That is not the way the curriculum has developed over the lifetimes of the current faculty, at least. Courses and topics are viewed as inherently belonging to an undergraduate program, and many have never questioned the purpose of certain topics or courses. One participant lamented about this circumstance, saying, "It is hard for outsiders (or even insiders, come to think of it) to believe that there are parts of our curriculum whose purpose we do not exactly know."

Others agreed: "I am not convinced that our curriculum is designed from goals to courses." In fact, the evidence is that it is the reverse, from courses to goals, if goals are even articulated. Another added, "I have been troubled for many years with the lack of systematic curriculum evaluation in our department. I once gathered the minutes of the committee meetings for a ten-year period looking for the rationale for our curricular decisions. Usually changes were due to young faculty trying to reform our curriculum to conform with one that they had seen elsewhere."

One participant outlined the general task by saying, "We need to identify what we want to teach—including the non-content ideas such as mathematical maturity, problem-solving ability, and ability to write and understand proofs."

Some who have begun the process of articulating goals have encountered new problems. One reported, "My feeling is that the department here, after two years of discussion, is more polarized over philosophy than at any previous time. Mathematics faculty are not equipped to articulate assessment goals. The MAA has to step forward." Others pointed to other problems by observing, "We have too little knowledge about how well our service courses prepare students." "Any type of student achievement assessment needs to take student characteristics into account, especially when it is used to compare programs."

Some who had worked on goals articulation had suggestions. One such suggestion was to look at all the final examinations a student takes over the degree program as a start. "I have often told beginning teachers who are trying to state goals to use the final examinations as a first approximation. Of course, we have many goals which do not appear on these exams. The lamentable but frequently asked question, 'Will this be on the exam?' is particularly nettlesome when the answer is 'no.' We all know this happens frequently."

Another participant offered general goals: "One of our major goals is preparing our students to go out and continue learning on their own, whether in an academic setting or not. Another is enabling them to organize and communicate what they do know."

A recent AAUP Committee report on mandated assessment was not optimistic about being able to measure the goals in a student's major field of study. The following statement on assessment in a major field of study is taken from that report [3, p. 38]:

> Most faculty members agree on the importance of assessing systematically a student's competence in the major, as shown by the multiplicity of forms of assessment that many departments employ. Yet even in this disciplinary context the range of possible student options after graduation makes it unlikely that an externally-mandated assessment instrument would do anything more than gauge the lowest common vocational denominator. The major is properly regarded as a vehicle for deepening the student's independent research and study skills, and thus standardized assessment of achievement in the major field raises precisely the same objections as it does in general education.

> Learning for its own end—for the purpose of developing breadth, intellectual rigor, and habits of independent inquiry—is still central to the educational enterprise; it is also one of the least measurable of activities. Whereas professional curricula are already shaped by external agencies, such as professional accrediting bodies and licensing boards, the liberal arts by contrast are far more vulnerable to intrusive mandates from other quarters; for example, the governors' report professes to find evidence of program decline "particularly in the humanities." To be sure, even in the liberal arts a student's accomplishments in the major can be measured with relative objectivity by admission procedures at the graduate and professional level that include GRE scores as one of the bases for judgment. But a student majoring in English may wish to pursue a career in editing, publishing, journalism, or arts administration (to name only a few); a political science major may have in mind a career in state or local government or in the State Department. Either of them may have chosen his or her major simply out of curiosity, or perhaps out of a desire to be a well-educated citizen before going on to law school or taking over the family business.

> For these reasons we suggest that the success of a program in the major field of study is best evaluated not by an additional layer of state-imposed assessment but by placement and career satisfaction of the student as he or she enters the world of work. Whereas imposed assessment measurements will at best—and rightly—attract faculty cynicism and at worst lead to "teaching to the test," no responsible faculty member will ignore the kinds of informed evaluation of a program available through a candid interchange with a graduating senior or recent graduate.

Worries About Assessment

Assessment is not just the average score of your majors on a multiple choice test.
—*David Lutzer, College of William and Mary, 1991*

Two questions are confronted on campuses where mandated assessment is a fact or being considered:

- How can the mathematics faculty be motivated to take the lead in assessment of the major?
- How can assessment be harmoniously embedded into an academic program?

These questions prompted a bit of debate among the participants. Faculty members are suspicious. How will the assessment data be used? Will honest assessments be rewarded or punished? Will assessments be consistent across departments? What is in it for faculty members except more work? The rewards system needs changing, according to one participant who suggested, "To motivate the faculty to take the lead we should redefine 'research' to include scholarly activities related to the teaching of mathematics and reward that research (when it is of high quality) as well as we reward the publication of new theorems."

One supporter of assessment countered the skeptics:

> I cannot imagine a department that would say that 'we have educational goals, but we aren't really concerned about whether our students achieve them' ...Assessment should have consequences. A properly done assessment can be used to focus both departmental and administration attention on what must be done to improve a program. Self-knowledge on the part of a department should lead to self-improvement. Whether or not it does is the true test of a local administration's commitment to assessment.

Use of Assessment Data

Many faculty members believe that assessment is a prelude to cutting programs and that those who mandate assessment and publication of the results will not be content unless programs are cut. This is believed to be the only way that legislators and administrators can prove to a skeptical public that they are doing their jobs. This is comparable to the belief that the only way that the public will believe that faculty evaluations are useful is that some faculty members are dismissed as a consequence. "Colleagues express concerns about how the administration will use such assessments. The fear is that they will be misused. Somebody is always ready to dictate what programs are of value and which don't deserve to be supported."

Generally, participants with the less experience in assessment voiced more negative feelings:

> The typical faculty member does not believe that there are any benefits to externally mandated assessments. The problem is that the purposes of assessment are not clear and different purposes are confused.

> One has to overcome the faculty perception that assessment is a no-win situation.

> I have serious reservations about assessment of entire colleges and universities since it is external, and thus invites abuse, misuse of information, and use of very shallow measuring instruments. In addition, it is very difficult, if not impossible, to compare entire colleges and schools with a simple assessment procedure.

> It is clear that any assessment statements must address at least two different kinds of mathematics departments—departments at undergraduate colleges and departments at universities. Changes are going to be easier to make at colleges, simply because of the size of the departments involved and the fact that all of the instruction is undergraduate.

These comments were offset by participants with experience that had generated support for assessment. One reasoned,

> When the state began its mandated assessment program, the university responded with a locally controlled program of assessment which was largely ignored by faculty in most departments (including mathematics). We have learned very little about our curriculum or instruction from the decade of data gathering which has taken place. ...Nevertheless, I am a believer in the need for

some assessment. The article by Grant Wiggins [9] was most interesting and inspirational, and I would look there for where we should be headed.

In addition to general suspicions, there are some specific worries about assessment, especially externally mandated assessment. These include having assessment shape the curriculum, infringement on academic freedom, compromising the balance between academic integrity and democratic direction, and the costs of assessment programs.

Shaping the Curriculum

The most obvious and alarming worry is movement toward what has been called the positivist approach to the curriculum—only aim to teach those things that are objectively assessable, or "teaching only to the test." In this way, assessment limits the curriculum negatively. One participant was certain: "The danger with assessment is exactly that it shapes the curriculum and the way in which courses are taught." Another added, "I fear that if assessment is mandated from the outside that the mathematics faculty will not be the ones to set the goals." Yet another braved the wrath of his colleagues, saying, "I support assessment shaping course content under the right circumstances. (I have colleagues that would string me up for that statement.) However, I believe the goals have to be set first."

In 1990 Ernst Benjamin [4] wrote in the *Chronicle for Higher Education* that "State-mandated assessment is dangerous not because evaluation is inappropriate, but because the requirement that universities demonstrate their quality in politically acceptable or popular terms—unmediated by the expertise of an accrediting body or the systematic procedures of a governing board—deprives universities of the safeguards that insure a balance between academic expertise and democratic direction."

One participant disagreed with Benjamin's thesis. "I see this as a red herring. If a department sets its own goals, evaluates its own success at achieving those goals, and uses that evaluation as a basis for proposing new strategies to reach the goals (and perhaps proposes changes in those goals), what could be the possible harm? The disagreement over this question seems to be growing out of a too narrow definition of assessment ..."

Academic Freedom

How might assessment, budgetary incentives, and public scrutiny threaten academic freedom and faculty control over the curriculum? The AAUP worries about threats to academic freedom and has stated a position on "Mandated Assessment of Educational Outcomes" in the November/December 1990 issue of *Academe* [3]. The following are extracts from the "protections for the role of the faculty and for reasonable institutional autonomy" as recommended in the AAUP Committee report [3, p. 40]:

- The faculty should have primary responsibility for establishing the criteria for the assessment and the methods for implementing it.
- The assessment should focus on specific, institutionally determined goals and objectives, and the resulting data should be regarded as relevant primarily to that purpose.
- If externally mandated assessment is to be linked to strategic planning or program review, the potential consequences of that assessment for such planning and review should be clearly stated in advance, and the results should be considered as one of several factors to be taken into account in budgetary and programmatic planning.

- The assessment process should not rely on merely one instrument, and both the process and the method of gathering information for the purposes of assessment should themselves be subject to continuing faculty review.
- The agency that mandates assessment should bear the staffing and other associated costs of any assessment procedure beyond those incurred as a result of faculty action.
- Under no circumstances should an assessment procedure be used to evaluate individual students or faculty members, nor should any procedure serve as a basis for comparing disparate academic units or institutions.

There was some disagreement among the participants as to whether assessment, externally mandated or not, was a serious threat to academic freedom. If mathematics department faculty members controlled the assessment—both implementation and use of results—there would be little threat. Many faculty members are already accustomed to departmental syllabi and even department-wide examinations.

One said, "There is a threat to academic freedom if and only if assessment is hopelessly misused." Another countered, "Assessment and the publication of the results can be a major threat to the balance between academic freedom and faculty control over the curriculum. This is particularly true when the assessment is simplistic and the results are used to make major budgetary decisions. For example, if our test scores don't 'improve' each time, we may lose a substantial amount of funding. This leads to the search to improving scores rather than educating our students."

The new wrinkle in the current assessment movement of using student performance to evaluate programs and teaching raised serious concerns among the participants. Said one, "Implementing the use of student performances to evaluate teaching is very complicated. The true test of teaching is how much the students learn and how they can use what they have learned." What are the risks in using student performance to judge teaching practice, teacher performance, curriculum, programs, institutions, and systems?

The risks are many, indeed, according to one participant. "Look at the situation in athletics where performance is judged by student performance. The recruitment scandals are just the tip of the iceberg, and the coaches are never asked to teach 'service' courses!" If student performances are used in program and teaching evaluation, then there will be the same incentives to 'cheat' as are present in intercollegiate athletics. Some elaborate control mechanism like that imposed by the intercollegiate athletic associations may become necessary.

Costs

The costs of an effective assessment program are generally believed to be large, both in faculty time and in materials. However, those perceptions about materials costs may be inflated. At one institution, over the last four years, fifteen arts and sciences departments carried out reportedly effective assessment programs for an average annual cost of $3,000 each. That covered the cost of two external reviewers per department, senior and alumni surveys, special testing, and copying of portfolio materials.

Nevertheless, participants voiced their concern about large costs. "The costs of a truly effective assessment program are probably enormous. In these days of financial shortages I doubt if we would be able to bear the costs of even a modestly effective program at campus levels without a reduction in the quality or quantity of education we provide."

Concern about the size of the costs are not all the worries. Where the money will come from also worries some people. An official of one state government was quoted as saying that assessment program could be funded by the savings from eliminating inadequate programs.

Examples of Programs

What a liberal education is about—and what assessment must be about—is learning the standards of rational inquiry and knowledge production.

—Grant Wiggins, 1990

According to the discussions of this focus group, there are not very many exemplary programs of assessment now, especially not of programs of assessment in the undergraduate mathematics major. When asked if there are exemplary programs, one participant with particular insight responded cautiously: "One hopes that there are. But, as a resident of a campus which has acquired a reputation for leadership in assessment of general education due to ten years of effort by some pretty capable and devoted people, I would conclude that there are no such examples. If our program has gained as much credibility as it seems to with as little impact as I can find there must be very little high quality competition."

Nevertheless, there are some programs that appear successful. Two participants described the programs at their institutions as follows.

In the department we have chosen to look at our program objectives and to develop means of assessing each of them. The result will be to require each senior to develop a portfolio containing (1) a report of their senior project which might be a variety of things from a presentation, paper, course project, or teaching unit, (2) a collection of graded problems and proofs from upper-division courses, (3) a completed attitude instrument, and (4) scores from a standardized examination.

We have what is still called a comprehensive examination requirement in the major, which has evolved from a written examination of all the major courses taken into a requirement for independent study (with a faculty advisor) and oral and written presentation of the results. Most of us think of this as more of a capstone experience, tying together a lot of loose threads and moving the student to the next stage of independence (usually), but it does give us a lot of non-quantified information about the strengths and weaknesses of our students as mathematical thinkers.

Portfolios of students' work are becoming a common part of assessment programs. These are generally viewed as snapshots of the students' work over a given program. The collection of a student's final examination papers would diverge a bit from the snapshot idea, but would provide a cross-sectional view of some goals of the program and how the student met those goals. However, not all goals are tested on examinations.

Course-embedded assessment is receiving increased attention. One major research institution, although reporting no significant activity in comprehensive assessment, reported institution-supported research projects in course-embedded assessment. Examples of areas of inquiry include acquisition of critical thinking skills as they relate to major concepts in a course, students' abilities to integrate content across topics, and effectiveness of student learning experiences such as collaborative learning, use of technology, and internship placements.

What Should the MAA Do?

What should the MAA do? Publish some guidelines! This request came up over and over again. There was a similar refrain in the late 1970s about placement and diagnostic testing. The MAA responded then through the Committee on Placement Examinations, now Committee on Testing, but never really gave the recipe that many people craved. That is probably how this new assessment effort will play out also. The situations at various institutions are so different that only very general guidelines can be recommended to apply to all.

Further, participants believed that the MAA should state clearly that assessment must be a departmentally based process of goal statement followed by evaluation of success in achieving goals, followed by a combined departmental/administration effort to improve vis-a-vis those goals. Members want to know what kind of tools and approaches are being used effectively. They want a thoughtful discussion of what assessment can mean to a department, and they especially want samples of learning goals statements (maybe for specific mathematics courses) from departments with assessment programs.

One experienced assessment coordinator stated:

> After two years experience in helping departments with assessment, I think there are three things that would be useful: (1) sample statements of student learning goals—the most difficult thing to produce—or an MAA statement on student learning objectives in undergraduate mathematics; (2) some methods for determining student achievement; and (3) a rationale for assessment that would persuade MAA members that evaluating undergraduate education by focusing on what is happening to the students (specifically, what students are learning in relationship to what we would like them to learn) is a good idea.

Others want the MAA to serve the role that Ernst Benjamin [4] calls the mediation of the expertise of an accrediting agency or the systematization of a governing board's procedures in controlling the use of assessment. Samples of requests in this direction follow.

- "The MAA should make a clear statement about the purposes of assessment and the proper use of the data generated."
- "The MAA should formulate and publish a statement that details the ways in which external evaluations should be structured and conducted, the ways in which the results are presented and interpreted, and the range of acceptable actions based on the results of a mandated external assessment."
- "The MAA should be developing and announcing policy that talks about the forms of assessment of departments and programs that are valid and acceptable, about the ways in which data should be gathered during an assessment and the ways in which those data should be analyzed and synthesized, about the qualifications of the persons who conduct the assessment and interpret its results, and about the ways in which changes, based on assessment results, should be made so as to strengthen and not damage existing programs and departments. That is, to delineate the purposes and means of valid assessments clearly."

Finally, one participant brought the problem home, saying, "A number of our committee members have talked about the MAA doing various grand and wonderful things with respect to assessment. From where I sit, we are that MAA committee. At this point there is certainly no consensus about what the terminology and problems are, let alone what the solutions can be."

Focus Group Participants

aaupeb@gwuvm.gwu.edu — ERNST BENJAMIN, *Amer. Assoc. of University Professors.*

bushaw@wsuvm1.bitnet — DONALD BUSHAW, *Washington State University.*

comers@citadel.bitnet — STEPHEN COMER, *The Citadel.*

pa24948@utkvm1.utk.edu — HENRY FRANDSEN, *University of Tennessee.*

harvey@math.wisc.edu — JOHN HARVEY, *University of Wisconsin.*

djlutz@wmvm1.bitnet — DAVID LUTZER, *College of William and Mary.*

bmadison@uafsysb.uark.edu — BERNARD MADISON, Moderator, *Univ. of Arkansas.*

richard_millman@csusm.edu — RICHARD MILLMAN, *Calif. State Univ., Santa Barbara.*

wm@virginia.edu — NED MOOMAW, *University of Virginia.*

cpeltier@bach.helios.nd.edu — CHARLES PELTIER, *Saint Mary's College.*

rhoades@iubacs.edu — BILLY RHOADES, *Indiana University.*

romberg@vms.macc.wisc.edu — THOMAS ROMBERG, *University of Wisconsin.*

math19j@jetson.uh.edu — JAMES STEPP, *University of Houston.*

joswaffo@ilstu.bitnet — JANE SWAFFORD, *Illinois State University.*

References

1. "Time for Results." Report of the National Governors' Association Task Force on College Quality, August 1986.

2. Steen, Lynn Arthur (Ed.). *Challenges for College Mathematics: An Agenda for the Next Decade.* Report of a Joint Task Force of the Mathematical Association of America and the Association of American Colleges. *Focus*, November-December, 1990.

3. "Mandated Assessment of Educational Outcomes." *Academe*, November-December 1990, 34–40.

4. Benjamin, Ernst. "The Movement to Assess Students' Learning Will Institutionalize Mediocrity in Colleges." *Chronicle of Higher Education*, July 5, 1990.

5. Edgerton, Russell. "Assessment at Half Time." *Change*, September-October 1990, 4–5.

6. Hutchings, Pat and Marchese, Ted. "Watching Assessment: Questions, Stories, Prospects." *Change*, September-October 1990, 12–38.

7. Marchese, Ted. "A New Conversation about Undergraduate Teaching: An Interview with Professor Richard J. Light, convener of the Harvard Assessment Seminars." *AAHE Bulletin*, May 1990, 4–8.

8. Wiggins, Grant. "A True Test: Toward More Authentic and Equitable Assessment." *Phi Delta Kappan*, May 1989, 703–713.

9. Wiggins, Grant. "The Truth May Make You Free, But the Test May Keep You Imprisoned: Toward Assessment Worthy of the Liberal Arts." The AAHE Assessment Forum, 1990, 17–31.

Appendix: Toward Assessment Worthy of the Liberal Arts

The Truth May Make You Free, but the Test May Keep You Imprisoned

by Grant Wiggins, CONSULTANTS ON LEARNING, ASSESSMENT, AND SCHOOL STRUCTURE

I confess that I have been kicking myself for getting involved with this topic. The more I looked at the title I obligated myself to, the more nervous I became. For one thing, I do not know as much as I should about current efforts to assess higher education. For another, I do not think anybody in his or her right mind can address this topic intelligently in so little time. And third, I think that the problem specified in the title of my talk is an insoluble one. It confronts us with one of many inescapable dilemmas about the liberal arts: the freedom of thought to go where it will versus the apparent need for uniformity in the testing process. In other words, we are not going to "solve" the assessment problem in the liberal arts, now or later. We are going to negotiate it—painfully; we are going to have to deal with some frequent, uneasy compromises.

So what I intend to do is a bit more modest than perhaps it first seemed. My aim is not so much to lay out a complete vision, but to give you my sense of the subtle but profound shifts that would be required if we were going to be serious about assessing for a liberal arts education. Second, I am going to propose to you a set of principles that we might call upon when liberal education is jeopardized—as it always will be—by an overly utilitarian or vocational view of teaching and learning. I think of these principles as mere first cuts, but perhaps they can hold you in good stead on a rainy day. Third, I will offer what I hope are some provocative and useful illustrations of alternative forms of assessment that befit the liberal arts. Many of the examples happen to come from the K–12 arena but nonetheless apply to your situations as well. I would encourage you, therefore, to resist a common nasty, little habit. If I should make reference to a fifth grade teacher's example, try not to be snooty about it. It is harder than you think to resist the feeling, and harder still to develop the almost anthropological mindset that enables one to find insight into one's own teaching from very different places in the system.

Thoughtless Mastery

Let us begin thinking about dilemmas in education by returning to the first known assessor in the liberal arts. I am thinking, of course, of Socrates, the Socrates of the dialogues of Plato, where we regularly see those who either appear to be or profess to be competent put to the "test" of question, answer, and—especially—sustained and engaged conversation. (The dialogues themselves, of course, are filled with dilemmas. Many of them are left unresolved: a reminder of how these arts are meant to lead to questions, not answers—the little burrs that get under your saddle.) Socrates the assessor: he is certainly a strange one. He does not seem to have nice answer keys or scoring rubrics by his side. Yet I think that there is something to learn from thinking about assessment from a Socratic point of view.

I would like to view these issues through my favorite piece of literature, the dialogue called "Meno." Some of you no doubt know it. Meno, a brash young fellow, comes up to

Reprinted with permission from *The AAHE Assessment Forum*, 1990, pp. 17–31.

Socrates. The first line of the dialogue is, "Tell me, Socrates, how do we become virtuous?" In other words, he is asking how morality develops: through upbringing? moral education? by nature? Socrates responds in a very annoying and typically Socratic way. He says, "Well, I cannot answer that question. I do not even know what virtue is." Meno is clearly astonished to think that this could be possible, that a *bona fide*, certified sage does not know what everybody knows, namely, what it means to be good. But of course, after Meno makes the foolish mistake of venturing to tell Socrates what virtue is, Socrates proceeds to undress him two or three times.

Finally, in exasperation, Meno says a terribly revealing thing that goes to the heart of the distinction between conventional assessment done well and an assessment for the liberal arts. Meno says, "Well now, my dear Socrates, you are just what I have always heard before I met you. Always puzzled yourself and puzzling everyone else. And you seem to me to be a regular wizard. You bewitch me. You drown me in puzzles. Really and truly my soul is numb. And what to answer you I do not know." And here is the important part. "Yet I have a thousand times made long speeches about virtue before many a large audience. And good speeches too, as I thought. But I have not a word to say at all as to what it is."

Meno's comment (and indeed the progress of the whole dialogue) ironically reveals what so differentiates conventional academic mastery from excellence befitting the liberal artist. Meno is reduced to speechlessness, he thinks, because of the sophistry of Socrates' questions and analyses; the thoughtful reader knows, however, that Meno does not know what he is talking about. And yet Meno is a conventionally successful student. How do we know? Throughout the dialogue Meno is constantly dropping references—the ancient equivalent of footnotes—to all the famous people who say this and that about virtue, which he, of course, agrees with. And it is no doubt the case that Meno could be a successful speaker—effective, convincing. The point of the dialogue, of course, is that such rhetorical skill using borrowed ideas is not understanding; competent presentations are not sufficient. That is not what a liberal education is about.

What Socrates wants us to see—what Plato want us to see by the way in which the dialogue is written—is that the conventional view of education is actually quite dangerous. If one gets better and better at what one does, one is less and less likely to question what one knows. Meno has been a dutiful student. (We are also meant to know that his name is a pun: It is very close in Greek to the word for memory: menon–mnemon.) Meno is an effective memorizer, able to make effective speeches with references to famous people. Isn't that what too much of our assessment is already about? Don't we too often fail to assess whether the student can do anything more than borrow quotes, facts, and figures?

But we also know from history that the real Meno was a nasty fellow: clever, ruthless. We are meant to know that. Because there is ultimately a lesson to be learned about "control" over knowledge and the ends to which "mastery" is put. Liberal education can never co-exist happily with other, more "practical" views of education because a liberal education is about rooting out thoughtlessness—moral as well as intellectual thoughtlessness.

There is, alas, such a thing as "thoughtless mastery" (as I have elsewhere termed it) and our syllabi and assessments tend unwittingly to reinforce it. Many of our students are quite good at this thoughtless mastery; you all know it. You know those looks in class, those mouth-half-open looks, the eyes slightly glazed; when people are fairly attentive but the brain does not seem to be quite engaged; when, alas, their eyes only focus to check scores on other people's papers, and to press you for extra points here and there.

Paradoxically, many professions *require* unthinking mastery—and run the risk of an amoral technical approach to life. I think we forget this. I do not want the pilot who flew me to Washington to be questioning his knowledge or his existence. Nor do I want my brain surgeon to be thinking about what virtue is. One of my passions is baseball, and I was recently reading George Will's new book called *Men At Work* on the craft of playing and managing major-league baseball. There is an odd but insightful phrase in it about this kind of thoughtless mastery that rings quite true. The good hitters talk about not thinking too much—that it is very dangerous to do so. Rather, what has to take over the hitter is something called "muscle memory"—a wonderful phrase for the kind of unthinking skill that we admire in athletes.

There is no reason, however, for colleges and universities to assume that their job is to promote unthinking mastery of others' ideas (while also abetting the other forms of thoughtlessness that too easily follow). Colleges are derelict, I think, in giving up the only sanctioned time when we have *a moral obligation to disturb students intellectually.* It is too easy nowadays, I think, to come to college and leave one's prejudices and deeper habits of mind and assumptions unexamined—and be left with the impression that assessment is merely another form of jumping through hoops or licensure in a technical trade.

Certainly we *say* we would like to see more "real" thinkers, and we bemoan doltish behavior in our students. I think we protest too much. Our testing and grading habits give us away. If you do not believe me, look how often students give us back precisely what we said or they read. On the other hand, you should not think that I mean rigor does not matter. That is part of the dilemma. The great mistake that has been made in school reform by many so-called progressives, and by much of the alternative schools movement, is to assume that to be liberated is to be liberated from discipline. That is a mistake, and it is one reason why alternative school people end up shooting themselves in the foot: because they produce a lot of free spirits who are not always very capable. If I had to choose, I might go with the alternative schools, but it is a bad choice and it shows that we have not negotiated the dilemma in K–12.

So we have to think about rigor. We have to think about alternative assessment as more than just engaging students better, which it invariably does (you know this if you have done simulations, case studies, portfolios, or dramatic presentations with your students). We need more than engaging activities. We need truly standard-setting and standard-revealing assessments. Or as psychologist Lauren Resnick puts it: What we assess is what we value. We get what we assess, and if we don't assess it, we won't get it. True about rigor, but also true about the intellectual virtues.

Some of you know, if you have read some of the things that I have written on alternative assessment, that one of my definitions of authentic assessment is that it is "composed of tasks that we value." It is not a proxy. It is not an efficient system to shake out a grade. Efficiency and merely technical validity as the aims of assessment will *undermine* liberal education. Rather, the test should reveal something not only about the student but about the tasks and virtues at the heart of the subject—its standards. But it is damn hard to design tasks to meet those criteria. It is very easy to score for efficiency and to look at what is easy to score rather than what is essential.

Three Dilemmas

Let me cite three other dilemmas before giving you some principles and examples of how we might think about assessment that would do justice to the liberal arts. The first dilemma, confronting you more often as a teacher the higher up you get in the system, is whether to stress students' mastery of the ideas of others or mastery over their own emerging ideas. In fact, we do believe that it is important for students first to control subject matter and to acquire skill within the discipline before they get "creative." Or to paraphrase Thomas Kuhn, one must have complete control over the existing "paradigm" if dramatic new paradigms or original thoughts are to occur.

Whatever Kuhn's merits as a historian and philosopher of science, I think he is dead wrong about education. I think it is terribly important that would-be liberal artists immerse themselves, from the word go, in questioning the paradigm *as* they learn it: They should study it, poke it, prod it, and not wait until they have mastered it—because you can have a *long* wait. And many of your bright and able minds are likely to drop out mentally or physically because they cannot wait that long. Conversely, the ones that stick around may be more dutiful than thoughtful.

Inevitably, if we first demand control over the subject matter in its entirety, we run a moral as well as an intellectual risk. We run the risk of letting the student believe that authority and authoritative answers matter more than inquiry. We may well end up convincing students that "Knowledge" is somehow something other than the result of personal inquiries built upon questions like theirs. And in fact, many students do believe that: There is "Knowledge" over here and never the twain shall meet.

A second way to put the dilemma is more classic: useful versus useless knowledge. There is an important sense in which the liberal arts *are* useless, summed up in that little comment supposedly made by Euclid 2,000 years ago when someone complained that geometry was not good for very much. He said, well, give him three drachmas if he has to get some usefulness out of studying it. But there is a more important truth in this desire. It is not at all clear that this unending inquisitiveness and poking over, under, and around knowledge is useful. Indeed, I can tell you from working with adolescents for so many years (prone to outbursts of honesty and not feeling the need to appear like eager apprentices), that many of them regard it as profoundly useless. On the other hand, we must ourselves keep clear the distinction between "useful" (or "relevant") and "meaningful." Students are *not* entitled to usefulness in a liberal education, but they *are* entitled to a meaningful encounter with essential ideas. We often disappoint—either by pursuing ideas that are *too* relevant but transitory, or by being insensitive to their need for provocations, not packages of pre-digested "knowledge," to chew on.

Third, we have to recognize that the urge to shun the liberal arts may have a great deal to do with the essential urge to feel competent. People go to school, it seems to me, indirectly to feel good about themselves. They want to develop competence because they want to develop confidence—or is it the reverse? The trouble with a liberal education is that it does not satisfy that need at all. It is unpleasant. It is disturbing. Many people drop out mentally and become hyper-competent because they cannot deal with the ambiguity and uncertainty that is the hallmark of the liberal arts.

Well, then, suppose I am right about this. Suppose we are in danger of treating assessment in higher education—as we are now increasingly in danger of treating assessment in lower education—as certification that a student possesses sanctioned knowledge. Where would we look for effective alternative strategies? How can we highlight the liberal arts side of the dilemma? What principles might guide us in designing assessments for the liberal artist in training?

Let me offer you ten tentative principles.

Principle 1

The heart of the liberal enterprise is not a mastery of orthodoxy but learning to justify one's opinions. Because the modern university has its roots in the Middle Ages and in religious training, it is built upon an irresolvable tension between orthodoxy and the promotion of inquiry. We tend to forget that. To this day, it seems to me, we still lean pretty heavily on the orthodoxy side: Up until the graduate experience, students have first to demonstrate their control over other people's knowledge. Yet we would be wise to begin our reforms from the perspective of the ultimate educational experience with which we are all familiar: the dissertation and oral in defense of a thesis. We should think of all assessment as designed primarily to give students an opportunity to *justify* opinions that are being developed as they explore subject matter.

This implies that one of the most important things that we can do in assessment is to examine the students' response to our follow-up questions and probes of their ideas. It implies, for instance, in assigning a paper and evaluating it, that the student should have to formulate a response, to which we then respond as part of the formal assessment process, not as a voluntary exercise after the test is over or the paper done.

Taken to the limit, I would argue that one of the most important things that we can do with students is to assess them on their ability to punch holes in our own presentations. They have a right to demand from us justification of our point of view. That is what the liberal spirit is about. It sends a moral message that we are both, student and teacher, subservient to rational principle.

Principle 2

The second principle is that we really need to think of the student as an apprentice to the liberal arts. And like all apprentices, students should be required to recognize and produce quality work. They should not get out of our clutches until they have produced some genuinely high-quality work. Now, what do I mean by that? Well, it is really a subtle shift in thinking. We all expect quality as teachers, but I do not believe that we demand it.

For instance—and here is one of those sixth grade examples—there is a teacher in Louisville who in one of her first assignments to her social studies students demands that every student read a book and do a book report. Not a particularly interesting task, but what is fascinating is what she demands. She demands that the paper be perfect. She demands that the students not turn it in until it is. She demands that they seek out anyone and everyone who will help them make it perfect.

Well, needless to say, the kids freak out. Especially the bad ones who are convinced that they cannot produce quality work. To make a long story short, they do. Oh, we could quibble with the idea of a perfect paper, but the kids understand full well what is meant. They really do. It is quite something to see. They understand that they have to ratchet up the seriousness with which they work. That they cannot wait to find out the quality of the work they produced. That they have to produce the quality work *first*. Making a point that many of you know is now critical to the alternative assessment conception: Assessment and self-assessment must be intertwined if we are serious about empowering people. To demand quality is also to structure assessment so that the student does not merely have the opportunity to rehearse, revise, rethink, but is actually required and expected to do so.

One of my favorite assignments when I taught at Brown was to ask students for their final paper to rewrite their first paper, based on all they had since learned or thought. A number of the upper-classmen told me that it was the most important event in their four years at Brown. They were astonished to see how their thinking had changed. They were astonished to discover how sloppy that early work seemed to them in retrospect. In short, they were learning about quality.

Further, they were learning about thinking, that thinking does not stand still and should not. Demanding quality, in other words—and this is part of the shift in thinking that is required—means we begin to focus our assessment on what Aristotle called the intellectual virtues. Does the student display craftsmanship, perseverance, tolerance of ambiguity, empathy when everyone else is critical, a critical stance when everyone else is empathetic? Can the student, without prodding, re-think and revise a paper or point of view? A liberal arts education is ultimately about those intellectual virtues. When all of the knowledge has faded away, when all of the cramming has been forgotten, if those intellectual dispositions do not remain, we have failed.

Now, some people get very squeamish about assessing things like perseverance, style, craftsmanship, love of precision. I do not. If we value it, we should assess it. That does not mean that we are arbitrary. That does not mean that we are subjective. Yet, we have to worry about validity and reliability. In fact, what I think it means to assess habits of mind is not to directly score them at all. But rather to devise tasks that *require* the habits we value.

My metaphor for this is "Outward Bound." Assessment should be like intellectual Outward Bound. It should reveal to the student what we value as traits in them by the virtues required to accomplish the task at hand. It should not be possible to do an end-run around those habits; students who can get A's by missing class and cramming are telling you something about the failings of your assessment system.

Sometimes it is as subtle a shift as sending the message day in and day out that quality matters and you are held accountable for quality. One of my favorite little tricks in that regard comes from Uri Treisman at Berkeley and his work with minority mathematics students. He demands that every piece of work the students hand in be initialed by another student; students get a grade both for their own paper and for the paper on which they sign off. This sends a message loud and clear that quality matters, that you are personally responsible for quality, and that it is in your interest to find out about quality *before* hearing from the authority. Quality control is about avoiding poor performance before it happens.

Principle 3

This leads directly to Principle 3, a point familiar to many of you who have been at this kind of work, but one that cannot be made often enough. A liberal arts assessment system has to be based on known, clear, public, non-arbitrary standards, and criteria. There is no conceivable way for the student to be empowered and to become a masterful liberal artist if the criteria and standards are not known in advance. The student is kept fundamentally off-balance, intellectually and morally, if the professor has a secret test and secret scoring criteria.

Consider the performance world, as opposed to the academic world, and how much easier it is for performers to be successful because of this very basic fact. The test is known from day one. The music, the script, the rules of debate, the rules of the game are known: genuine mastery in the performance arena means internalizing public criteria and standards until they become one's own. Unfortunately, in education, especially in higher education, there is a vestige of our medieval past, when tests were a bit of mystery and novices had to divine things. I was disappointed to learn when I was a teaching assistant at Harvard that most undergraduates are still not allowed to see their blue books. And then somebody told me that at Oxford and Cambridge they burn them.

I think this is an unfortunate and deadly tradition. It is also a legacy of tests used as gatekeepers, not as equitable vehicles designed for displaying all that a student knows. Most people in this room, I suspect, would say it is the *student's* responsibility to figure things out, to respond to the test as the test demands, and to produce quality work on *our* terms. I am not convinced of that. Why isn't the university required to meet students half-way and give them a chance to reveal their strengths and play from their strengths? It would be as simple as giving people the option of alternative forms of doing the same assignment.

But I think it runs much deeper. We are still using testing as a sorting and categorizing system. And elitism should not be confused with meritocracy. Our most common habit in scoring and grading, namely scoring on a curve, is unjustifiable in my view. Its sole purpose is to exaggerate difference rather than reveal strength. It makes our life easier and it relieves us of justifying the grades and scores that we give. It is needlessly debilitating—as opposed to a challenge that we can rise to when we know, understand, and appreciate the criteria.

Of course, many of you know the solution. Scoring rubrics, model papers, video-taped model performances, anything that can give students an insight into, allow them to enter the field and acquire its standards, by seeing exemplary performance *before* they do their work. I do not know why in the world we keep such matters a secret. It is cuckoo—and dysfunctional.

Principle 4

It follows that what a liberal education is about—and what assessment must be about—is learning the standards of rational inquiry and knowledge production. And this implies that self-assessment is a critical and early part of assessment. Now, many of you know about Alverno's use of self-assessment and it has been borrowed by many of us. I just want to give you one of my favorite Alverno examples because I think it illustrates so well different ways of thinking about this.

One of Alverno's goals for students is competency in oral communication. Early on, a student must give a video-taped talk and so one's first hunch is, oh, well, you are going to assess the talk. No. After the student gives the first talk and it is video-taped, the student is assessed on the accuracy of the self-assessment of the video-taped talk. That is a fundamental shift in point of view. If we want people to gain control of important habits of mind and standards, then they have to know first of all how to view those things accurately and apply criteria to their own work, and not always depend upon another person to do that.

It is also a basic lesson in habit development. You have to know what you are *supposed* to be doing before you can do it. And that knowledge is crucial in making you stick with it and believing that it is possible. Otherwise I do not think any of us would quit smoking or lose weight. It suggests as a practical corollary that no major piece of work should get turned in without some self-criticism attached to it. And that self-criticism should be assessed for *its* accuracy.

Principle 5

Most education, it seems to me, treats the student as a would-be "learned spectator" rather than a would-be "intellectual performer." The student must metaphorically sit in the bleachers while others, mostly professors and writers of textbooks, perform. This arrangement takes us back to the idea that competency involves just remembering and applying what others say. It has dangerous consequences because it induces intellectual passivity. In an education for a would-be performer the student would experience the same "tests" that face the expert in the field—having to find and clarify problems, conduct research, justify one's opinion in some public setting—all while using (other people's) knowledge in the service of one's own opinion.

Let me give you a couple of my favorite examples of this. One of the finest classes that I have ever observed at any level was at a high school in Portland, Maine, where a veteran teacher offered a Russian history course. The entire syllabus consisted of a series of chronological biographies. It was, however, the *student's* job to become each person, in turn, in two senses: through a ten-minute talk, and then a simulation. After four or five students had presented their talks (and been assessed by other students on their talks), they had a Steve Allen "Meeting of the Minds" press conference which was chaired by the teacher; the "journalists" were the other students. Each party scored the other for its performance.

Now, I do not know about you, but I have sat through a lot of dreary reports. These were not dreary. In fact, they were as interesting and informative as any reports I had ever heard. I went up to the teacher and said, golly, how did you get them to do that? He said, well, it was very simple. There were only two criteria by which they were going to be judged and they were (a) whether the talk was accurate, and (b) whether it was interesting. This was real performing and using knowledge.

Principle 6

This one follows from Principle 5. A liberal artist, if he or she has "made it," is somebody who has a style. Somebody whose intellectual "voice" is natural and clearly theirs. Read the turgid prose that we receive and you know that we are failing to develop style, voice, and point of view. (Read our own writing in journals ...) Students are convinced we want the party line, and that the quality and insight possible in compelling prose is not necessary. It is an option.

There would be a simple way to get at this. After writing a lengthy research paper with all the requisite footnotes and bibliographical information, the student could be asked to turn the paper into a one-page piece to be delivered, in an engaging and insightful way, to an audience of lay-persons. But it is not just an aesthetic issue, this business of style. It is a question of one's inner voice. One's intuition. The seed of a new idea that is easily crushed if it is not allowed to be heard. All of these are related to the idea of conscience, and, of course, it is no coincidence that Socrates talked about his little voice.

It is easy for students in American universities to lose that little voice. But that little voice is not just a "personal" voice irrelevant to "academic" accomplishment. It is the voice of common sense. It is the voice that can turn around and question the importance of what one has just spent two months working on. It is the little voice that says, ahh, come on, is this really *that* important? Or it is the little voice that says, you know, there is probably another way to look at this. It is the little voice that says, I have a feeling that there is something behind what the professor is saying, but I do not know enough to really pursue it so I will not. It is the little voice that most of us do not hear in our students unless we ask for it. An assessment should ask for it.

Such assessing need not be difficult. I saw an English teacher do it. In using peer editing, he told his students that they should reject and turn back any paper that was boring or slap-dash—and mark the exact spot in the paper where they began to lose interest. Nothing sends a message faster to students about writing and its purpose and quality. Nothing sends a message quicker that technical compliance with criteria is not always of primary importance.

There is another point to be made about voice and style. The thing that is so ghastly about academic prose is that one really does sense that it is not meant for an audience. And, of course, sometimes it is not. It seems to me that if we are serious about empowering students, we must get them to worry about audience in a deeper way. We must demand that their work be *effective*. We must demand that it actually reach the audience and accomplish its intended purpose. There is nothing more foolish, in my view, than saying, "Write a persuasive essay" without the students having to persuade anybody of anything. So let us set up situations in which the student *has* to persuade readers, or at least get judged by an audience on more than just accuracy. Even Socrates knew, in the clash of Reason and Rhetoric, that teaching had to be not merely truthful but effective.

Principle 7

Too often in assessment we worry about whether students have learned what we taught. This is sensible, of course. But let me take an unorthodox position. Such a view of assessment, taken to extremes, is incompatible with the liberal arts. One important purpose of those "arts that would make us free" is to enable us to criticize sanctioned ideas, not merely re-tell what was taught.

A less confrontational way to make the point is to remind ourselves that it is the astute questioner, not the technically correct answerer, who symbolizes the liberal artist. The philosopher Gadamer (with an explicit homage to our friend Socrates) argued that it is the dominant opinion that threatens thinking, not ignorance. Ensuring that students have the capacity to keep questions alive in the face of peer pressure, conventional wisdom, and the habit of their own convictions is what the liberal arts must always be about.

Admittedly, *some* knowledge is required to ask good questions and pursue the answers we receive. But if we are honest about this we will admit that the kind of exhaustive expertise we typically expect in students is over-kill. After all, children are wonderful and persistent questioners; recall the wisdom of H.C. Andersen's *The Emperor's New Clothes*. Indeed, academics are invariably prone to making the mistake Gilbert Ryle called the Cartesian fallacy: assuming that "knowing that" must *always* precede and serve as a condition for "knowing how." No person who creates knowledge or uses knowledge to put bread on the table would ever be guilty of this fallacy. All apprentices or would-be performers learn on the job, yet as teachers we tend to over-teach or "front load" knowledge. So a good pedagogical rule of thumb would be: teach the minimum necessary to get the students asking questions that will lead to your more subtle goals.

We would do well, then, to think of our task as introducing the student to cycles of question–answer–question and not just question–answer—with one aim of a course being to make the student rather than the professor the ultimate initiator of the cycle. To postpone developing students' ability to ask important questions in the name of "mastery" is to jeopardize their intellect. Good judgment and aggressive thinking will atrophy if they must be endlessly postponed while professors profess. In any event, the most important "performance" in the liberal arts is to initiate and sustain good question asking.

A very mundane point about testing can be made out of this esoteric argument. We rarely assess students on their ability to ask good questions. Indeed, we rarely teach them a repertoire of question-asking strategies for investigating essential ideas and issues. It should become obvious to students through the demands of the course and our assessment strategies that question-asking is central. Too often, however, our assessments send the message that mastery of the "given" is the exclusive aim, and that question-asking is not a masterable skill but a spontaneous urge.

Principle 8

This principle follows from Principle 7. The aim of the liberal arts is to explore limits— the boundaries of ideas, theories, and systems. To paint the starkest picture of the difference between a "liberal" and a "non-liberal" view of the disciplines, therefore, we might see our task as teaching and assessing the ability to gauge the strengths and weaknesses of every major notion we teach—be it a theorem in mathematics, a hypothesis in science, or literary theory in English. We need to know whether students can see the strengths and weaknesses of "paradigms." This would include the limits of a theory not only within a subject but across disciplines, as when we apply the rules of physical science to the human sciences.

There is no novelty in this idea. I am invoking a notion about the liberal arts developed thirty and more years ago by Joseph Schwab at Chicago. He termed such a view of education the art of "eclectic," and I encourage you to return to his essays for numerous suggestions on how to help students explore the merits of sanctioned truths.

I fear that we no longer know how to teach science as a liberal art in this sense. When we make science merely abstruse and technical, we make it increasingly unlikely that non-scientists will profit from studying science enough to support intelligent science policy as adults. And we encourage science students to become too technical and insufficiently critical. I really think that the first years of study in college (and certainly throughout secondary school) have less to do with "mastering" science and more to do with orthodox algorithms— learning metaphysics instead of physics, sanctioned truth vs. the unstable results yielded by methods and questions that *transcend* the current results.

I know this weakness in our science students first-hand from my high school teaching days. My students did not understand, for example, that error is inherent in science and not merely the fault of immature students or poor equipment. (Many believe that when the "big boys and girls" do their measuring, the results are exact.) Nor did many of them realize that words like *gravity* or *atom* do not correspond to visible "things" to be seen directly.

The point can be made another way. We still do a poor job of teaching and assessing the student's grasp of the history of important ideas. I know of no other method by which inappropriately sacred truths can be more effectively demystified. What questions was Newton, then Einstein, trying to answer? What did the first drafts of a history text look like, and why were they revised? To ask questions like these is to open up a new and exciting world for students. To be smug about our knowledge and to distance ourselves from "crude" and out-dated theory is to ensure that we repeat the mistakes of our smug and parochial elders.

Consider the history of geometry, the very idea of which strikes many people as an oxymoron. Many college students are utterly unaware of the problems that forced Euclid to develop an awkward parallel postulate (which was instantly decried by his colleagues). So much for self-evident truths, that glib line found in superficial views of Greek mathematics! Further, most students are unaware that non-Euclidean geometries can be proven to be as logically sound as Euclid's; they have no idea how that result transformed epistemology for all fields of study.

The consequence of our failure to reveal to students the history of important ideas is two-fold. For one, students easily end up assuming that axioms, laws, postulates, theories, and systems are immutable, even though common sense and history say otherwise. The second result follows from the first and does lasting harm to intellectual courage in all but our feistiest students. Students never grasp that "knowledge" is the product of "thinking"— thinking that was as lively, unfinished, and sometimes as inchoate as their own. One major reason for the intellectual poverty in this country is that most students either become convinced they are incapable of being intellectual, or they are uninterested in it if it involves only arcane expertise in a narrowly framed subject.

Some practical implications for assessment? First, we should require students to keep notebooks of reflections on coursework, their increasing knowledge, and important changes of mind about that knowledge. Second, we should assess this work as part of the grade. I did so for many years and found it to be the most important and revealing aspect of the students' work. I also learned a lot about how their thinking evolved in a way that improved the courses I taught. Third, even the most technical training should ask students to do critical research into the origins of the ideas being learned so that students can gain greater perspective on their work. If we fail to do this, whether out of habit or rationalization that there is no time for such reflective work, we risk producing a batch of thoughtless students.

Principle 9

Principle 9 extends the moral implications of Principle 8. When we encourage narrow, unchecked expertise, we may unwittingly induce students to be dishonest about their ignorance.

I am not even talking about the more heinous crime of cheating, something we know is all too common. Rather, I am talking about the moral obligation of the liberal artist to emulate Socrates' trademark: his cheerful admission of ignorance. Alas, our students rarely admit their ignorance. One of our primary tasks should be to elicit the admission and not penalize it. But the student's willingness to take such a risk depends upon our doing so. It is then, as the "Meno" reminds us, that mutual inquiry and dialogue become possible because we are placed on equal moral footing as thinkers. More pointedly, our inclination to "profess" is always in danger of closing off the doors through which our students can enter the liberal conversation without excessive self-deprecation. So many of our students preface a wonderful idea by saying, "I know this sounds stupid, but ..."

Let our assessments, therefore, routinely encourage students to distinguish between what they do and do not know with conviction. Let us design scoring systems for papers that heavily penalize mere slickness and feigned control over a complex subject, and greatly reward honest admissions of ignorance or confusion. Or, let us go the next step and ask students to write a second paper in which they criticize the first one.

Principle 10

The tenth and last Principle extends the point. Intellectual honesty is just one aspect of self-knowledge and the absence of self-deception. One of my favorite notions was put forward by Leo Slizard in talking about how to assess doctoral candidates. He argued that students should be assessed on how precisely and well they knew their strengths and limitations—and that it was a mistake to err greatly in *either* direction.

I am not arguing for professors to become counselors. I am arguing for them to improve students' ability to self-assess, and to make sure that accurate self-assessment is more than a pleasant exercise. It is an essential tool for ensuring that students have neither excessive nor insufficient pride in their work, either of which closes off further intellectual challenges and rewards.

The inherent danger of scholarship is not so much error as blind spots in our knowledge—hidden by the increasingly narrowed focus of our work and the isolation that can then breed worse arrogance. Excessive pride leads us not only to ignore or paper over our doubts but more subtly to be deceived about the uniqueness and worth of our ideas—we forget that it was a conversation in the coffee shop or reading an article that sparked the idea. A few collaborative assessment tasks, with some reflection on the role of each contributor, would provide useful perspective for everyone.

It follows that we should assess class discussions more than we do. We again fail to assess what we value when we make it possible for students to learn everything from only listening to us and doing the reading. I and many others have developed some good material for assessment (and self-assessment) of class discussions, and I encourage you to develop some methods of your own.

Which of course brings us back to Socrates. What the casual reader of Plato always fails to grasp—including some overly analytic philosophers, I might add—is that the dialogues invariable are about character, not "theories" of virtue, knowledge, or piety. The twists and turns of dialogue, the sparring with Sophists or young know-it-alls—ultimately all this is meant to show that character flaws, not cognitive defects, impede the quest for knowledge. It is one's *attitude* toward knowledge that ultimately determines whether one will be a liberal artist or merely proficient.

As Socrates repeatedly reminds us, we must love wisdom so much that we question our knowledge, even our pet ideas if need be. By extension, the more we gain confidence in our ideas, the more we must become vigilant about finding knowledge in unexpected places—and let others who seem incapable of it teach us something, as our students often do.

It is not a canon—of ideas or books—that defines the liberal arts, but a set of very hard-won virtues. Like all sophisticated dispositions, these liberal habits are typically only revealed when they are challenged. It is only when peer pressure is greatest, be it in the classroom with students or at conferences with our peers, that we learn who has the power to keep questions alive. The liberal arts, properly speaking, do not make you free; they *keep* you free. Wisdom—Socrates knew—reveals itself when persistent inquiry is threatened: externally by custom and "oh, *everyone* knows …;" and internally by the tendency to rationalize our own habits, beliefs, and fears.

How much do students really love to learn, to persist, to passionately attack a problem or task? How willing are they, like the great Indian potters of New Mexico, to watch many of their half-baked ideas explode, and start anew? How willing are they to go beyond being merely dutiful, perfunctory, or long-winded? Let us assess such things, just as the good coach does when he or she benches the talented player who "dogs" it.

We are then quite properly assessing not skill but intellectual character. It is to our detriment and the detriment of the liberal arts if we feel squeamish about saying and doing so. Let us "test" students in the same way that the mountain "tests" the climber—through challenges designed to evoke whether the proper virtues are present. And if not present, the quality of the resultant work should seem so inadequate to the *student* that little need be said in the way of "feedback."

Let our assessments be built upon the distinction between wisdom and knowledge, then. Too subjective? Unfair? Not to those who have the master's eyes, ears, and sense of smell—tact, in the old and unfortunately lost sense of that word. For these traits are as tangible as any fact, and more important to the student's welfare in the long run. It is not the student's errors that matter but the student's response to error; it is not "thoroughness" in a novice's work that reveals understanding but awareness of the dilemmas, compromises, and uncertainties under the arguments one is willing to stand on.

If our testing encourages smug or thoughtless mastery—and it does—we undermine the liberal arts. If our assessment systems induce timidity, cockiness, or crass calculations about grades and the relevance of today's assingment, we undermine the liberal arts. If our assessments value correctness more than insight and honesty, we undermine the liberal arts. If our assessments value ease of scoring more than revealing to students the errors or tasks that matter most, we undermine the liberal arts. Let us ensure, above all else, that our tests do just what Socrates' tests were meant to do: help us—and our students—to distinguish the genuine from the sham authority, the sophists from the wise. Then we will have assessments that are worthy of our aims.

Untying the Mind's Knot

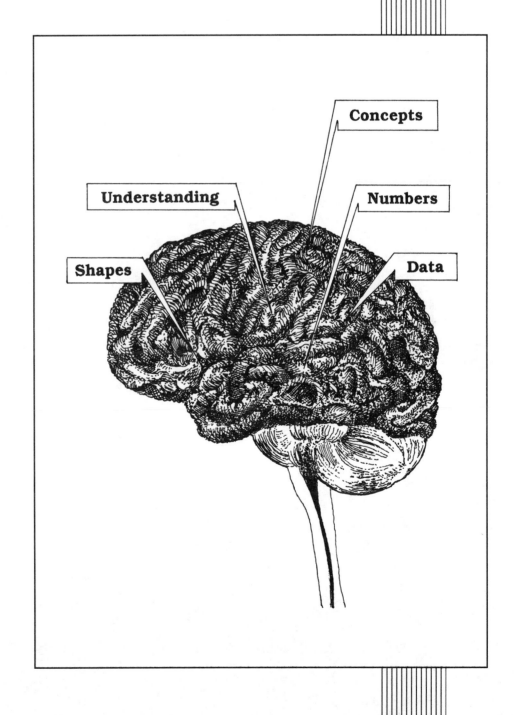

Concepts

Understanding

Numbers

Shapes

Data

Untying the Mind's Knot

Barry A. Cipra

NORTHFIELD, MINNESOTA

> *Listen and learn,*
> *for what I shall now say*
> *will be a gift of lofty consequence.*
>
> > —*The Paradiso, Canto VII, 23–24 (Ciardi).*

> *Opinions too soon formed often deflect*
> *man's thinking from the truth into gross error,*
> *in which his pride then binds his intellect.*
>
> > —*Canto XIII, 118–120.*

Not unlike the poet Dante, undergraduate mathematics has lost its way in a dark woods. Many who set out as teachers find themselves wandering aimlessly through the groves of academe, waiting for a Virgil or a Euclid to guide them to the light. Some can't see the forest for the trees. Some seek salvation in the roots, others in the canopy. Some see nothing but danger lurking, and a steep, accelerating descent into the nested circles of Hell and Damnation.

Among the many voices in the wilderness are some crying "Research, we need more research!"

Others retort, "Like hell we do!"

What's all the shouting about? Is Research to be our Beatrice—or is the call for careful study another Siren's song, sweetly luring scholars to their doom? The answer depends on who you ask.

This report will examine some of the issues and controversies surrounding research in collegiate-level mathematics education. It is based on interviews with a number of mathematicians and mathematics educators who hold a variety of opinions on the subject. The purpose here is not to survey or update the latest studies, but rather to bring into focus the central concerns of those working in the area and of those who wonder if it's worthwhile.

> *You will certainly come to know your view*
> *is steeped in falsehood if you listen well*
> *to the counter-arguments I shall offer you.*
>
> > —*Canto II, 61–63.*

A Diversity of Definitions

To start with a seemingly innocuous question: What is research in collegiate-level mathematics education? That gets quickly to at least one source of the cacophony of opinions, because there does not seem to be any general agreement on the meaning of the term.

The spectrum of interpretations as to what such research is, not to mention what it's good for, makes the fable of the blind men and the elephant look like a model of consensus and agreement.

"The term means different things to different members of different communities," says Jim Leitzel, a professor of mathematics at Ohio State University currently at the MAA headquarters in Washington, DC. The diversity of definitions makes dialogue difficult.

One way to view such research is as pure inquiry into the processes of human understanding: The learning of mathematics poses numerous fundamental questions in cognitive science; the job of the mathematics education researcher is to follow those questions wherever they may lead. If the journey happens to bring something useful back to the classroom, then so much the better. But that's not necessarily the reason for doing the research.

That point of view should be familiar to mathematicians, who smile knowingly at G.H. Hardy's peculiar boast, "I have never done anything 'useful.'" But when it comes to educational research, mathematicians tend to think in terms of their own needs. They become the client, with a client's demands for service. From that point of view, mathematicians see educational research in purely practical terms.

"For the most part the mathematical community, when it comes to research on mathematics education, is a user community. That is, the generic mathematician says, no different from the high school teacher, tell me what I can do in my classroom on Monday," says Alan Schoenfeld of the University of California at Berkeley. "It is appropriate for mathematicians to ask that of educators, but it is not appropriate for that to be what defines mathematics education research."

Schoenfeld points to three broad categories of research. The first he calls social and epistemological "engineering." This work is based on the view of mathematics as a human activity—a view mathematicians are likely to stick with even after computers have taken over completely. The operative pedagogical assumption is that students develop this view of mathematics only if they experience it by engaging in the "give-and-take of mathematical sense-making," rather than the customary note-taking approach in which the student copies a neatly packaged version of the subject. What goes on in this type of setting and how to make it work are questions of interest to a number of researchers.

The second category Schoenfeld calls "product-oriented" work. This work is closely related to curriculum development and includes many such efforts, but it differs by placing an emphasis on foundational aspects and research. The difference could be likened to the difference between an applied mathematician who's interested in the general conditions under which a particular numerical technique will produce an approximate solution to a differential equation and an engineer who just wants the damn thing solved.

Schoenfeld's third category is "basic" research into cognition. Like basic research in any subject, this work often has no obvious or immediate applicability, but presumably will exert a long-term influence. Much of Schoenfeld's own work falls into this category. In one project, for example, Schoenfeld and colleagues analyzed seven hours of videotape showing a single student interacting with a tutor and a computer program. "It goes without saying that I see this work as being deeply connected to some of my more 'applied' mathematical work," says Schoenfeld, "but I must stress that on its own terms this is basic work pursued independent of potential applications: our huge investment in analyzing seven hours of videotape was aimed at yielding a better understanding of mathematical thinking and learning processes."

Some feel the time is ripe for the development of theories of understanding. Others are not convinced. "There's a continuing distrust, or skepticism, about any theoretical explanation of how the brain processes information concerning mathematical concepts," says Ramesh Gangolli of the University of Washington. On the other hand, Gangolli notes, there's a growing willingness on the part of mathematicians to take a chance trying out some of the educational practices suggested by results from cognitive science. "I think there are some encouraging signs that professional mathematicians are now coming to regard some of the results of cognitive work as useful," he says.

Schoenfeld acknowledges that "research methods are not well established at this point." However, he says, the situation is much better than it was. "About 10 or 15 years ago, the field finally came to the realization that the methods that were pretty much standard were essentially worthless," he says. Those methods were "scientistic" misapplications of statistical methods such as comparison studies and factor analyses which, while appropriate for agricultural studies, were out of place in the classroom. "The problem is that human behaviors—and humans generally speaking—are just a lot more complex than fields of corn."

The field has gotten away from that, Schoenfeld says. Instead, "we started asking: Can we build some models of thought processes? Can we get a handle on what those things actually are and talk about the way that cognitive structures grow and change? That's a jargon way of saying, can we really describe learning in a precise way?" The change in thinking has left the field "in pretty much an exploratory phase." Nevertheless, Schoenfeld maintains, "I think we've come close to identifying the salient aspects of mathematical thinking. We at least know what it isn't, which is the reductive view that everybody had of it 15 or 20 years ago, that the mathematics you know is the sum of the mathematical facts and procedures you know. We have a much richer view [now]."

> *I approve what has emerged thus far, but now*
> *it is time you should explain what you believe,*
> *and from what source it comes to you, and how.*
>
> *—Canto XXIV, 121–123.*

500 Papers

Traditionally, mathematics education has concerned itself with children before they get to college, and especially with youngsters in the elementary grades. Researchers look into how children learn to count, what happens—or goes wrong—when they add and subtract, how various societal problems affect achievement in mathematics, and a vast number of other questions.

Research along the same lines at the collegiate level is less common and has a much lower profile. Most members of the mathematics community, even when they have an abiding interest in educational issues, are unfamiliar with the research literature. To the extent that they've heard rumors of it, many are suspicious.

The amount of research in collegiate-level mathematics education is relatively small by modern standards. A dedicated scholar (or masochist) could conceivably read the entire extant literature, provided he or she had access to the various sources. Joanne Rossi Becker and Barbara Pence at San Jose State University surveyed the most prominent sources for the years 1975–1989, and compiled a list of approximately 500 published manuscripts. (They chose 1975 as a starting date because there was already a bibliography, compiled by Marilyn N. Suydam, for 1900–1974.) Becker and Pence also identified over 750 dissertations involving college-age populations in the same period.

Becker and Pence classify research under the general headings of student learning, methodology, equity issues, academic preparation for college, and teacher education. There is necessarily overlap among and between them. Student learning includes research into learning styles, problem solving, concept formation and misconceptions, and specifically into the learning of algebra and calculus and the impact of computer technology. Under methodology, they cite studies of effective teaching, the use of tests and assignments, and experiments with cooperative learning.

"Equity issues" includes both gender and minority issues. It also includes studies of "affective variables"—e.g., "math anxiety" studies. Research on college preparation—much of which is now done by the colleges themselves—focuses on predictors of success (such as the use of placement exams) and the development of remedial courses. Teacher education refers to research on how well teacher education programs do at preparing pre- and in-service teachers, an issue of increasing importance in light of the "sweeping changes" called for by such documents as the NCTM *Standards* and the MAA *A Call for Change*.

Compared to the thousands of papers published every year in traditional research areas of both mathematics and mathematics education, the several hundred papers identified by Becker and Pence constitute a manageable amount of material. There are signs, though, that this will change. The growing educational reform movement—itself still a fairly young enterprise in spite of its long history—has led many teachers to reconsider the basic tenets of their craft, and some are inclined to do so in systematic fashion.

"Things are very different from the way they were three or four years ago," says Ed Dubinsky at Purdue University. "There are a number of people who are now actively involved in research in mathematics education at the post-secondary level."

Dubinsky is part of that crowd. "I've been doing research for about six or seven years, maybe a little longer if you count the time that I took when I started reading other people's stuff." He began with a study of Piaget, the Swiss psychologist whose studies of children have had a profound influence on educational theory. "I've tried to make myself something of a specialist in the ideas that Piaget had about how people learn, and to see if those ideas can be extended to the post-secondary level," Dubinsky explains. He has also spearheaded the effort to establish educational research as a legitimate professional activity for members of the mathematics community.

Research in mathematics education, Dubinsky asserts, has an appropriate place in the world of mathematics. "In this community of 45,000 people, there is a significant number who are hired by mathematics departments for their interest in education," he says. Those numbers indicate "a pent-up potential for sharp growth" in the area of educational research, he adds.

It's an uphill battle, though. Education researchers are faced with a mathematics community that is generally skeptical, occasionally hostile, and usually uninterested. There are lingering doubts about the quality of research in mathematics education. There is a communication gap separating researchers whose native tongue is educationese from a potential audience schooled in the language of mathematics—a gap that some fear, and others hope, is unbridgeable. And, as always, there is the ogre of priorities: Are the results of research in mathematics education worth an investment of resources?

> *Thus [Beatrice] began: "You dull your own perceptions*
> *with false imaginings and do not grasp*
> *what would be clear but for your preconceptions."*
>
> —*Canto I, 88–90.*

Quality Time

By far the biggest bone of contention in the whole discussion of research in collegiate-level mathematics education is over the question of quality: Can the educational research establishment convince the mathematics community that its standards are up to snuff? There are some for whom the answer is a resounding "No." Others are willing to take a wait-and-see approach. Still others defend the standards of the discipline as entirely adequate as is, albeit inadequately enforced.

The dispute stems partly from fundamental differences between mathematical and educational research. The nature of the work is different, and the standards for one do not apply to the other. In mathematics, while research is ultimately judged on its significance and depth, there is a first filter of logical (as opposed to political) correctness: Mathematicians prove theorems, and a proof is either valid or it isn't. (This glosses over many shades of gray, of course—or should we say "obviously"? For example, what one person sees as a full-blown proof others may read as the skimpiest of sketches. Also, mathematicians are amazingly forgiving of faulty proofs or even false theorems if the work has been led astray by deep new insights.) There's no corresponding abstract canon of rigor in educational research.

Instead, the criteria for quality in educational research have a more social cast to them. In particular, new research is judged in relation to previous studies—new work is expected to build on what others have found. (That's not to say it can't contradict earlier findings. It just shouldn't add apples to oranges.) One of the main criteria is the presence of a theoretical framework. Another is the presence of an established methodology.

These are admittedly fuzzy criteria, calling for considerable judgment on the part of people in the field. It's a paradigm that many mathematicians are clearly uncomfortable with, and some reject outright. Herb Wilf, for one. As editor of the *American Mathematical Monthly* from 1987 to 1991, Wilf, a professor of mathematics at the University of Pennsylvania, has been involved in discussions of whether or not the MAA should sponsor a journal devoted to educational research (of which more later). Of one batch of papers which were submitted as "good examples of the craft," Wilf says, "My own personal opinion was that very few of them—perhaps one or two—met normal standards for research that have been around in the Western world since the time of Newton."

"Most of the papers—almost all of them–were anecdotal in nature," Wilf goes on to say. That's not necessarily bad, it's just not research, he explains. "Personally I value anecdotal stuff. If an intelligent person tells you something that works and something that didn't work, then you're smarter than you used to be. So that's not so bad. But for some reason it wants to dress itself up as research in undergraduate education." In sum, he says, "I saw some stuff that definitely is in the category of good, intelligent, hardworking people communicating their experiences, and I saw almost nothing of what I would call research—carefully designed experiments with carefully regulated class sizes and comparisons of results, and so on. I saw almost none of that."

Joan Ferrini-Mundy of the University of New Hampshire agrees that many studies submitted as research are "hopelessly anecdotal," but disagrees on the criteria that distinguish valid research. "You look for hypotheses, for example, that are coming out of other work," she explains. "You look for some connection to something else, rather than just someone who decided they'd take a close look at something without worrying about what other people have already seen. You look for some kind of theoretical framework that's going to guide the questions, so that when they get in and start to look around they have something of a framework in which to interpret what they're seeing. And you look for systematic study of whatever it is that's being questioned—some kind of methodology that undergirds what's being done."

Those criteria can be met even when what's being looked at are interviews with students or examples of student writing or "some kind of data that feels anecdotal," Mundy says. Nevertheless, "it's a pretty fine line. It rests a lot, I think, on how much people are willing to trust the insights of the people doing the work."

What clouds the issue, everyone admits, is the presence of a great deal of research that's of low quality by anyone's standards. "The general feeling among mathematicians that there are a large number of papers that are of really low quality is a valid concern," says Dubinsky. Adds Schoenfeld: "The prevailing attitude is still one of cynicism, and I'd say a fair percentage of it is justified." But it's a bad idea, they argue, to throw out the baby with the bath water.

"It's very similar to an experience I had in the early 1960s, when I was starting out and did my research in general linear topological spaces, in which there was a similar high percentage of low-quality papers put out," says Dubinsky. "It was the sort of stuff that you could take a concept and write down the definition and find another context in which that definition was logically consistent, and go ahead and prove the same old theorems using the same old examples and counterexamples in a new context, and churn out the papers." As a result, Dubinsky recalls, other mathematicians developed a low opinion of the field. But "the appropriate response was not to throw the field out. The appropriate response was for some good people to do some really good stuff, and set the standards, and for people to develop the attitude that we're not going to treat as high-quality work things that are not high quality, but that there are things that are high quality, and we're going to encourage that."

The same holds for educational research, Dubinsky maintains. "The appropriate response is not to say there's too much junk in mathematics education research so let's just ignore the whole thing. I think the appropriate response is to recognize that there is a fair amount of very good stuff, and stuff that could be very useful—first of all, research of a high quality with its intellectual value as such, and in addition things that could be very useful to the practicing college teacher—and to try to find ways to develop standards that help us distinguish between those and to find ways to encourage the good stuff and discourage the weak stuff."

Then one of them came forward and spoke thus:
"We are ready, all of us, and await your pleasure
that you may take from us what makes you joyous."

—Canto VIII, 31–33.

Anybody Here Speak Jargon?

Would it help to have a journal devoted to research in collegiate-level mathematics education? Dubinsky and others argue that it would. Such a journal would help set standards for work in the field, and it would help focus attention on some of the issues in undergraduate mathematics education. Moreover, its existence would attract more researchers to work in the area and help establish educational research as a viable part of the world of mathematics.

But mathematicians are not only skeptical of the standards for quality in educational research, they are also put off by what they see as impenetrable jargon in the literature. How do you expect us to make use of this stuff, they ask, when we can't even make sense of it?

Researchers respond with a mixture of agreement and rebuke. On the one hand, they agree that there's more than enough jargon in their trade and that reading educational research papers is not a particularly pleasant job. "There is nothing more boring than reading good educational research," observes Gila Hanna of the University of Toronto. Hanna should know: she edits the journal *Educational Studies in Mathematics*, which rejects, she says, roughly 80 percent of the papers submitted.

On the other hand, educational researchers point out that they are not necessarily writing for an audience of mathematicians. Every field develops a technical terminology that's often opaque to outsiders. Mathematics certainly has an argot of its own. A mathematician decrying the jargon of educational research is analogous to an engineer complaining how hard it is to find useful information in the *Journal of Differential Equations*.

(One can only speculate on what a mathematician would find readable in a paper on educational research. Perhaps something like the following: Let L be a lecture given to a class C. Let $s \in C$ be a student, and let $U_L(s)$ be student s's understanding of lecture L. We will show that for almost all $s \in C$, $U_L(s) = 0\dots$.)

Nevertheless, it is incumbent on educational researchers to communicate with mathematicians. "There are two things going on," says Schoenfeld. "On the one hand, mathematics educators need to defend the work that's research *qua* research on its merits. On the other hand, mathematics educators need to take seriously the fact that there is a user community out there, that mathematicians will look for things to be stated in ways that are informative and useful for them. If there's a research finding that might be turned into something useful in curricular terms, the researcher can't just say, 'Here's the finding, mathematics community. Go ahead and do it.'"

At the same time, educational researchers must be careful not to oversell themselves— and mathematicians need to understand what research can and cannot provide. "Nobody is going to come across in mathematics education and say all of a sudden, 'We have a teaching method that's going to have students suddenly learn twice as much,'" notes Schoenfeld. "Actually many people do say that—and they're all charlatans."

"You don't do a single research study and say, 'Well, we've finally solved this problem, now we know the answer,'" says Ferrini-Mundy. "The best you'll get is something that confirms or extends somebody else's study. What you have are building blocks that come together after a while to point you in a direction." The process is not quick. "There's a strong expectation that educational research should give you answers that you can somehow use tomorrow," Ferrini-Mundy notes. "That expects too much of it."

For that matter, even when there are clear implications for classroom practice, it's far from clear that mathematicians will pay attention to the results of research. That's not just due to lingering skepticism. Nor is it merely a matter of making the results readable and available. The final, potentially insurmountable obstacle is simply inertia—or, to use a less obviously pejorative term, tradition. The recommendations of research could well fall on deaf ears. If educational researchers have discovered anything in observing classroom lectures, it's that telling people what to do, no matter how clearly you spell out the algorithm, is not a particularly effective way of getting the message across. On top of which, teachers are going to ask the same question they so dislike hearing from their students: Is this going to be on the test? How's it going to affect my tenure and promotion, how's it going to affect my status in the profession?

If that's the case, then should educational research be a priority of the mathematics community? Not if it detracts from the main effort, which is to get mathematics departments to take teaching seriously, argues Peter Hilton of the State University of New York at Binghamton. It would be "utterly foolhardy" to tell them to do so by concerning themselves with educational research, Hilton says. "We have to concentrate on the difficult problems, and not turn to the easy ones," he adds.

Hilton and others argue that top priority should be given to creative teaching and innovative curriculum development. Educational researchers don't disagree on the goal, but counter that the effort will best be served if it is well informed by the findings of sound research. Indeed, encouraging innovation in the absence of research smacks of irresponsibility: What if the pharmaceutical industry abandoned research in favor of innovative drug design?

> *I see now that your mind, thought upon thought,*
> *is all entangled, and that it awaits*
> *most eagerly the untying of the knot.*
>
> —*Canto VII, 52–54.*

Second-hand Smoke

Over the long run, the main contribution of educational research may be to change the way mathematicians think about teaching—in a sense, to substitute one set of preconceptions for another. If the research is carefully done, and if it's properly presented, then presumably the new preconceptions will be closer to correct than those they displace. For example, some current research indicates that some of the difficulties students have with calculus are traceable not to poor algebra skills, but rather to conceptual difficulties dealing with functions. It could be a noteworthy advance in the discipline if professors thought in those terms when they prepared their lectures. (It could be an even bigger advance if they dispensed with lecturing altogether, but let's not go into that.)

Schoenfeld likens this kind of effect to the effect that medical research has had on the health consciousness of the American public. The research itself is highly technical, and there's constant controversy, but over the course of time the more clear-cut, consistent findings have managed to sink in. Most notably, for example, the evidence linking tobacco with a whole slew of health problems has helped initiate changes in public attitudes toward smoking, to the point that smokers' "rights" are in serious jeopardy.

If the scientific crusade against smoking is an appropriate analogy, then educational researchers and reformers can expect a long, hard battle. The sheer weight of evidence won't tip any scales, but careful research can serve to inform the educational reform movement and help shape its future. The main benefit of research, say Hanna, is "the feeling that you operate from a position of knowledge rather than ignorance."

> *Open your mind to what I shall explain,*
> *then close around it, for it is no learning*
> *to understand what one does not retain.*
>
> —*Canto V, 40–42.*

Appendix A: Communicating Among Communities

A REPORT OF A CONFERENCE ON
Research in Collegiate Mathematical Education

The Mathematical Association of America (MAA), with support from the National Science Foundation, hosted an invitational conference on Research in Collegiate Mathematical Education in Washington, DC, November 8-10, 1991. Twenty-eight invited participants, representing the mathematics and mathematics education communities, candidly discussed issues relating to the growing interest among faculty and others concerning research in the teaching and learning of mathematics at the undergraduate level. The conference participants focused on four aspects of research in collegiate mathematics education:

- Communicating to college and university faculty the growing body of research in undergraduate mathematics education.
- Improving student learning by stimulating change in collegiate teaching based on the findings of this research.
- Encouraging high standards of research in undergraduate mathematics education.
- Supporting the increasing number of collegiate faculty who undertake research in undergraduate mathematics education.

It is no surprise that the views of those attending the conference varied considerably on these matters. There was general agreement on the urgency of seeking improvement in the teaching and learning of mathematics at the undergraduate level. A distinction was made

This Conference was supported by the National Science Foundation. Opinions expressed in this report are those of the authors, not necessarily those of the Foundation.

between general activity in support of improving undergraduate mathematics education and the role of research in the teaching and learning of mathematics at the undergraduate level.

There was some disagreement as to the role played by mathematics education research. Some felt that basic research in undergraduate mathematics education is an essential and continuing part of the process of change. Still others were skeptical that such research would have any effect at all, and cited the limited amount of persuasive evidence produced in a field which is perceived by many mathematicians to be jargon-laden. Another group argued that basic research is critically important but independent of attempts to improve undergraduate education. There was also a group who believed that there are some mathematicians in the teaching force who would not be motivated to change their habits regardless of how compelling the research results may be.

Those favoring more emphasis on mathematics education research argued that basic research in the learning and teaching of mathematics is an essential component of any endeavor that seeks to improve undergraduate mathematics. They cited as examples the many profound conceptual questions about students' understanding of function and the wide range of questions generated by the impact of technology for students' learning and doing of mathematics. They further argued that the growing number of individuals engaged in investigating these questions at the collegiate level requires scholarly support structures to ensure the vitality of their field.

Those opposed to placing more emphasis on mathematics education research argued that it is more critical to address current teaching practices and seek improvement. Furthermore, while many mathematicians are becoming more interested in improving the teaching of undergraduates, they do not have the inclination to learn the specialized vocabulary or understanding of techniques required to read or participate actively in research issues at the collegiate level. What would be beneficial, in this view, would be more access to survey or review articles, written in a more expository style, that convey the results of current research at the collegiate level.

Conference Statements and Recommendations

Many mathematicians tend to think of research in mathematics education as being concerned primarily with the improvement of teaching or the evaluation of some particular curricular innovation. For many attending the conference, the vast array of research topics being pursued by those working in pre-college and collegiate mathematics education was somewhat surprising. A sampling from that list includes:

- Students' understanding of the limit as a process and not a number;
- Stages of development in understanding the concept of function;
- Issues centering around problem solving;
- Issues addressing proof, logic, and reasoning;
- How students' learning styles may affect the context in which problems are posed;
- Students' use of visualization in "doing" mathematics;
- Translational difficulties as students move among graphical, symbolic, or numeric representations of ideas.

The well-established professional community doing research in school mathematics education has much to offer those investigating similar issues at the collegiate level. This is particularly so since many colleges and universities teach school-level mathematics in their entry-level courses and are confronting the need for more careful attention to these courses. Efforts to make connections between those pursuing education research at the pre-college and collegiate level are already under way.

Conferees generally agreed that a journal on research in collegiate mathematics education would be an important means of strengthening professional standards, of encouraging quality research, and of providing support for individuals doing this type of scholarly activity. However, to be effective in providing the scholarly support needed for workers in the field, such a journal would need to achieve sufficient stature to command the respect of the broad mathematical community—including both mathematicians and those doing research in mathematics education. Given the number of existing journals accepting papers in mathematics education research and the critical need for building awareness among the mathematical community as to the contribution that research can make to improve teaching and learning in mathematics at the undergraduate level, there was a clear feeling among the participants that launching such a journal at this time would not be the most appropriate first step. To deal with the present need for sharing the fruits of research in undergraduate mathematics education in as broad a way as possible, the following recommendations received consensus support:

Recommendation 1. *The Mathematical Association of America (MAA) and the American Mathematical Society (AMS), in cooperation with the National Council of Teachers of Mathematics (NCTM), should plan a series of annual volumes presenting exemplary research papers in collegiate mathematics education. These volumes would serve as precursors to the establishment of a journal.*

Conference participants believed that the reception of these annual volumes by the community would provide a measure of interest and of the need for further steps. To achieve maximum effectiveness of this recommendation, additional strategies will be needed to create awareness among the broader mathematical community of the issues being addressed by those working in this field of research. A full array of suggestions and strategies were noted by the conference participants. Many of these extend current means of communicating information at sectional, regional, and national meetings:

- More extensive use of mini-courses and contributed paper sessions;
- Developing appropriate panel presentations;
- Attempting to get topics on research in undergraduate mathematics on the agenda of Departmental Chairs meetings;
- Encouraging the program committee for national meetings to invite speakers to address these research topics;
- Encourage the professional societies to seek ways to develop summer faculty institutes for teaching and learning mathematics.

The final suggestion was for the societies to pursue vigorously the creation of a national network of centers for the teaching and learning of mathematics.

Recommendation 2. *Editors of MAA and AMS periodicals are encouraged to solicit substantial review or survey articles to appear simultaneously with publi-*

cation of the annual volumes to stimulate interest among mathematicians in issues addressed in these research volumes.

Most of the editors present supported the recommendation. The only concern expressed related to appropriateness for the mission of specific journals. In fact, some encouraged the submission of quality survey articles, written for an audience predominately consisting of mathematicians, for publication in their journals on an on-going basis.

Many participants noted that electronic linkages for sharing results among workers in the field are being established independently and through informal networks. It was strongly recommended that the MAA and AMS should investigate and develop as rapidly as possible appropriate mechanisms for more formal electronic exchange of information among individuals engaged in research on undergraduate mathematics education, and between members of this community of researchers and those who teach college and university mathematics. In particular, part of this investigation might include exploration of the possibility of establishing an electronic journal on undergraduate mathematics education. In the spirit of trying to enhance communication across the communities, the participants put forth two specific recommendations:

Recommendation 3. *Editors of journals and periodicals that are read by college and university mathematicians should provide regular information on research in undergraduate mathematics education through such means as brief "telegraphic" reviews of research papers and special survey articles or issue papers dealing with the application of educational research to the improvement of student learning.*

Recommendation 4. *The MAA and AMS, in cooperation with the NCTM, should plan special conferences or sessions at sectional, regional, or national meetings dealing with aspects of research in collegiate mathematics education. These activities should be designed to expand the interface between educational researchers and mathematicians.*

This recommendation further expands the current means of communicating information at sectional, regional, and national meetings of the professional associations. Other special meetings might be designed that would promote scholarly exchange among those actively engaged in research in collegiate mathematics education, for the purpose of setting a research agenda, or for the purpose of attracting newcomers to undertake investigation of issues important to the field.

There was considerable optimism about the perception that most mathematics departments are now more concerned about the teaching and learning of mathematics by their students, and that individual faculty frequently are discussing issues related to curriculum and teaching. There was concern, however, that some universities have addressed the issue by appointing special adjunct faculty to deal with instructional concerns of both students and faculty. This arrangement can lead to further isolation of the full-time faculty from activities in teaching and learning at the undergraduate level. Overall, there is much more faculty interest in looking for ways to improve student learning. Professional mathematics societies need to encourage this interest by identifying and publicizing in mainstream mathematics journals effective methods for stimulating and institutionalizing improved mathematics instruction. When appropriate, ties with the results of research findings in undergraduate mathematics education should be directly indicated.

To spur the continued development of department-based activity regarding issues of curriculum and teaching, professional societies should identify lists of prominent individuals who would be prepared to speak on issues of research in undergraduate mathematics education. Once compiled, this list should be sent to departments as suggestions for colloquium speakers and to program planning committees for sectional, regional, and national meetings. Before distribution, this list of speakers would be reviewed by the MAA Council on Education and the AMS Committee on Education. The appropriate committee at NCTM would also be consulted.

Data from the CBMS Survey *Statistical Abstract of Undergraduate Programs in the Mathematical Sciences and Computer Science 1990-1991* were cited to document the tremendous increase in the use of part-time faculty, particularly at two-year colleges. The Survey notes that there are far more part-time faculty than full-time faculty currently teaching at two-year colleges. Attempts to influence the teaching and learning of mathematics at these institutions must directly address this large instructional force. Conference participants expressed great concern about this becoming a serious barrier to creative change.

Recommendation 5. *The MAA and AMS, in conjunction with the American Mathematical Association of Two-Year Colleges (AMATYC), should undertake a study of the effects of the increasing reliance on part-time faculty for mathematics instruction, especially to determine in what ways part-time faculty may differ from full-time faculty in their approaches to teaching.*

A serious and frequently surfacing concern among conference participants was the need for research in collegiate mathematics education to become an "accepted" field of scholarly inquiry in mathematics departments. The field is relatively new and participation by creative, energetic individuals will be needed to enhance its vitality. There are some indications that as departments become more concerned about undergraduate education, they begin to broaden their definitions of scholarly activity. But this is by no means universal! In fact, several participants held the belief that mathematical research, and only that, should be the fundamental criterion for initial promotion and tenure in mathematics departments. Of course this must be carefully interpreted, for there are departments where individuals are specifically hired because of the contributions that they make in the field of mathematics education research and should be evaluated on that basis. However, given the growing number of individuals who are making significant contributions in the field of collegiate mathematics education, the issue cannot be ignored.

In addition, there are growing numbers of college and university faculty who are involved in highly creative curriculum projects or software development. These new directions for faculty, often recognized as valuable by the department (but sometimes not rewarded), require enormous amounts of time. If the definition of scholarship is broadened to include these types of activities, then there is still an issue of how contributions in the area can be adequately assessed. One clear suggestion was that the faculty need to write and publish results of their work. But even that task can be complex and suffers from differences in approach. There are those who pursue research using the methodologies inherent in the field of mathematics education and there are others who deal with innovative practice in the teaching and learning of mathematics where the resulting articles are more anecdotal in style. In fact, broad discussion of the evaluation of alternative forms of mathematical scholarship is needed by the community. Fortunately, the Joint Policy Board for Mathematics (JPBM) has recently established the Committee on Professional Values, Recognition, and Rewards. The conference directs the next recommendation to this committee:

> **Recommendation 6.** *The JPBM Committee on Professional Values, Recognition, and Rewards should seek to identify and disseminate effective evaluation and reward mechanisms that promote high standards in professional activities in mathematics education. In particular, the Committee's agenda should address the needs of those faculty whose professional work is devoted to research in mathematics education, as well as those whose work centers on curriculum development or educational practice.*

It will not be simple to implement these recommendations and suggestions. The final recommendation made by the conference is an effort to put in place a framework for monitoring progress on this report.

> **Recommendation 7.** *The MAA Ad-Hoc Committee on Research in Undergraduate Mathematics Education should take the necessary steps to request that it become a permanent, joint committee of MAA and AMS. When such a joint committee is established, NCTM and AMATYC should be asked to appoint liaison representatives to this committee. The charge to this new permanent committee should include the monitoring of progress on all the recommendations contained in this report.*

Those participating in the conference did not agree at every juncture, but they did reach consensus that to achieve any objectives at all will require the visible and active leadership of both the MAA and the AMS.

Conference Organization

There was little question that the Conference participants knew they were attending a "working conference." The first general session convened after dinner Friday evening, November 8. In all, there were six general sessions and two structured writing sessions. The general sessions were focused on the four aspects of the Conference noted in the introduction, and two sessions were devoted to responding to the work of the writing groups. At each of the two writing sessions, three groups dedicated themselves to developing sets of strategies for addressing the concerns raised in the general discussions. The three writing groups were each asked to discuss and make recommendations regarding a pre-determined set of questions. On the basis of individual interest, the Conference participants self-selected the writing group in which they would participate. What emerged was consensus on a variety of statements and recommendations. Many of these address the urgent need to communicate across the mathematics and mathematics education communities. The efforts to strengthen and enhance this communication is a full community task—not one that can be done alone by individuals or by any single professional association.

To the hard working participants of the Conference we express our profound thanks. Sincere appreciation is extended to the National Science Foundation without whose support the Conference would not have taken place. We also acknowledge the additional support for various conference activities provided by the American Mathematical Society.

Organizing Committee

DONALD J. ALBERS, *Associate Director for Publications and Programs, The Mathematical Association of America.*

ED DUBINSKY, *Purdue University.*

JAMES R.C. LEITZEL, *The Ohio State University, Chair.*

SAMUEL M. RANKIN III, *Associate Executive Director, The American Mathematical Society.*

LYNN A. STEEN, *St. Olaf College.*

Conference Participants

JOHN S. BRADLEY, *Associate Executive Director, American Mathematical Society.*

ALBERT A. CUOCO, *Woburn High School, Woburn, Massachusetts.*

JAMES DONALDSON, *Howard University.*

JOHN H. EWING, *Indiana University,* and *Editor, American Mathematical Monthly.*

BARBARA T. FAIRES, *Westminster College.*

JOAN FERRINI-MUNDY, *University of New Hampshire.*

DEBORAH TEPPER HAIMO, *University of Missouri–St. Louis,* and *President, The Mathematical Association of America.*

GILA HANNA, *Ontario Institute for Studies in Education,* and *Associate Editor, Educational Studies in Mathematics.*

JOHN HARVEY, *University of Wisconsin,* and *Editor, Journal of Technology in Mathematics.*

M. KATHLEEN HEID, *The Pennsylvania State University.*

RICHARD HERMAN, *University of Maryland* and *Joint Policy Board for Mathematics.*

PETER J. HILTON, *State University of New York, Binghamton.*

JAMES KAPUT, *University of Massachusetts, Dartmouth.*

LILLIAN MCDERMOTT, *Department of Physics, University of Washington, Seattle.*

GERALD PORTER, *University of Pennsylvania.*

RONALD ROSIER, *Georgetown University.*

CORA SADOSKY, *Howard University* and *Association for Women in Mathematics.*

ANNIE SELDEN, *Tennessee Technological University.*

RAY SHIFLETT, *Executive Director, Mathematical Sciences Education Board.*

MARTHA SIEGEL, *Towson State University,* and *Editor, Mathematics Magazine.*

GILBERT STRANG, *Massachusetts Institute of Technology,* and *Society for Industrial and Applied Mathematics.*

MARCIA P. SWARD, *Executive Director, The Mathematical Association of America.*

HARRIET WALTON, *Morehouse College.*

Guests Attending Several of the Sessions:

BARRY CIPRA, *Northfield, Minnesota.*

MARGARET COZZENS, *Program Officer, National Science Foundation.*

RAY HANNAPEL, *Program Director, National Science Foundation.*

JOAN LEITZEL, *Division Director, Materials Development, Research, and Informal Education, National Science Foundation.*

JAMES LIGHTBOURNE, *Program Officer, National Science Foundation.*

Appendix B: Reading List

This list of recent publications related to educational research at the undergraduate level was compiled by Ed Dubinsky of Purdue University in order to illustrate the nature and scope of current research.

Alibert, D., 1988. "Towards New Customs in the Classroom." *For the Learning of Mathematics*, 8(2) 31–35.

Artigue, M., 1987. "Ingenierie Didactique a propos D'Equations Differentielles." Proceedings of the 11th Annual Conference of the International Group for the Psychology of Mathematics Education (J.C. Bergeron and N. Herscovitz, Eds.), Montreal, 236–242.

Ayres, T.; Davis, G.; Dubinsky, Ed; Lewin, P., 1988. "Computer Experiences in Learning Composition of Functions." *Journal for Research in Mathematics Education*, 19(3) 246–259.

Blum, Werner and Niss, Mogens, 1991. "Applied Mathematical Problem Solving, Modelling, Applications, and Links to Other Subjects—State, Trends, and Issues in Mathematics Instruction." *Educational Studies in Mathematics*, 22(1) 37–68.

Breidenbach, D.; Dubinsky, Ed; Hawks, J.; Nichols, D., 1992. "Development of the Process Conception of Function." *Educational Studies in Mathematics*.

Chevallard, Yves, 1990. "On Mathematics Education and Culture: Critical Afterthoughts." *Educational Studies in Mathematics*, 21(1) 3–27.

Clement, J. "The Concept of Variation and Misconceptions in Cartesian Graphing." *Focus on Learning Problems in Mathematics*, 11(1) 77–87.

Confrey, J., 1991. "The Concept of Exponential Function." In L.P. Steffe (Ed.), *Epistemological Foundations of Mathematical Experience*. Springer-Verlag: New York.

Cornu, B., 1988. "Didactical Aspects of the Use of Computers For Teaching and Learning Mathematics." *Educational Computing in Mathematics*, North-Holland.

Freudenthal, Hans, 1982. "Variables and Functions." *Proceedings of the Workshop on Functions*. Enschede.

Goldenberg, P., 1988. "Mathematics, Metaphors, and Human Factors: Mathematical, Technical, and Pedagogical Challenges in the Educational Use of Graphical Representations of Functions." *Journal of Mathematical Behavior*, 7, 135–173.

Hanna, G., 1990. "Some Pedagogical Aspects of Proof." *Interchange*, 21(1) 6–13.

Heid, K., 1988. "Resequencing Skills and Concepts in Applied Calculus Using the Computer as a Tool." *Journal for Research in Mathematics*, 19(1) 3–25.

Kaput, J., 1987. "Representation and Mathematics." In C. Janvier (Ed.), *Problems of Representation in Mathematics Learning and Problem Solving*. Erlbaum: Hillsdale.

Leron, U., 1983. "Structuring Mathematical Proofs." *American Mathematical Monthly*, 90, 174–185.

Leron, U., 1985. "Heuristic Presentations: The Role of Structuring." *For the Learning of Mathematics*, 5(3) 7–13.

Monk, G.S. "Students' Understanding of a Function Given by a Physical Situation." In G. Harel and Ed Dubinsky (Eds.), *The Development of the Concept of Function*. Mathematical Association of America (forthcoming).

Orton, A., 1983. "Students' Understanding of Integration." *Educational Studies in Mathematics*, 14, 1–18.

Orton, A., 1983. "Students' Understanding of Differentiation." *Educational Studies in Mathematics*, 15, 235–250.

Owen, E. and Sweller, J., 1989. "Should Problem Solving Be Used as a Learning Device in Mathematics?" *Journal for Research in Mathematics Education*, 20, 322–328.

Schoenfeld, A., 1983. "Beyond the Purely Cognitive: Belief Systems, Social Cognitions, and Metacognitions as Driving Forces in Intellectual Performance." *Cognitive Science*, 7, 329–363.

Schoenfeld, A.H.; Smith III, J.P.; Arcavi A., 1990. "Learning—the Microgenetic Analysis of One Student's Understanding of a Complex Subject Matter Domain." In R. Glaser (Ed.), *Advances in Instructional Psychology*, 4. Erlbaum: Hillsdale.

Selden, A. and Selden, J., 1987. "Errors and Misconceptions in College Level Theorem Proving." In J.D. Novak (Ed.), *Proceedings of the Second International Seminar on Misconceptions and Educational Strategies in Science and Mathematics*, Cornell University.

Sfard, A., 1991. "On the Dual Nature of Mathematical Conceptions: Reflections on Processes and Objects as Different Sides of the Same Coin." *Educational Studies in Mathematics*, 22, 1–36.

Shaughnessy, J.M., 1977. "Misconceptions of Probability: An Experiment with a Small-Group, Activity-Based Model Building Approach to Introductory Probability at the College Level." *Educational Studies in Mathematics*, 8, 295–316.

Tall, D. and Vinner, S., 1981. "Concept Image and Concept Definition in Mathematics with Particular Reference to Limits and Continuity." *Educational Studies in Mathematics*, 12, 151–169.

Thompson, P., 1985. "Experience, Problem Solving and Learning Mathematics: Considerations in Developing Mathematics Curricula." In E. Silver (Ed.), *Teaching and Learning Mathematical Problem Solving*. Erlbaum: Hillsdale, 189–236.

Treisman, U., 1985. "A Survey of the Mathematics Performance of Black Students at the University of California, Berkeley." (Unpublished thesis available from the author at the Dana Center, Department of Mathematics, University of Texas at Austin, Austin, TX 78712-1082.)

Vinner, S., 1983. "Concept Definition, Concept Image and the Notion of Function." *International Journal of Mathematical Education in Science and Technology*, 14, 293–305.

Vinner, S. and Dreyfus, T., 1989. "Images and Definitions for the Concept of Function." *Journal for Research in Mathematics Education*, 20(4) 356–366.

Challenges for
College Mathematics
(An Agenda for the Next Decade)

Challenges for College Mathematics:

An Agenda for the Next Decade

This report was prepared by a Joint Task Force of the Mathematical Association of America (MAA) and the Association of American Colleges (AAC) as part of a national review of arts and sciences majors initiated by the AAC. The MAA was one of twelve learned societies contributing to this review. Each participating society convened a Task Force charged to address a common set of questions about purposes and practices in liberal arts majors; individual task forces further explored issues important in their particular fields.

The MAA-AAC Task Force comprised six individuals: JEROME A. GOLDSTEIN, *Tulane University;* ELEANOR GREEN JONES, *Norfolk State University;* DAVID LUTZER, *College of William and Mary;* LYNN ARTHUR STEEN, CHAIR, *St. Olaf College;* PHILIP URI TREISMAN, *University of California, Berkeley; and* ALAN C. TUCKER, *SUNY at Stony Brook.*

In 1991, AAC published a single-volume edition of all twelve learned society reports with a companion volume containing a separate report on "Liberal Learning and Arts and Science Majors." Inquiries about these two publications may be sent to Association of American Colleges, 1818 R Street, NW, Washington, DC 20009.

Generous funding for the project and dissemination of the reports was provided by the Fund for the Improvement of Postsecondary Education (FIPSE) and the Ford Foundation.

Preface

In response to a request from the Association of American Colleges (AAC), the Mathematical Association of America (MAA) convened a special Task Force to address a range of issues concerning the undergraduate major as a sequel to the 1985 AAC report *Integrity in the College Curriculum*. Whereas the 1985 *Integrity* study examined the effectiveness of general education, the new AAC study addresses the contribution that "study in depth" makes to liberal education. Each of the several AAC disciplinary task forces examined how study in depth relates to goals for the major, assurance of intellectual development, and relations with other fields.

The MAA-AAC Task Force operated in the context of two other simultaneous MAA reviews of the mathematics major. One was conducted by a subcommittee of CUPM, the Committee on the Undergraduate Program in Mathematics; the other by COMET, the Committee on the Mathematical Education of Teachers. Since these other committees are charged by MAA to provide specific advice to the mathematical community about requirements for the mathematics major, the MAA-AAC Task Force dealt only with broader questions that are central to the AAC study. Hence this report is not intended as a detailed statement about curricular content, but as a statement of issues and priorities that determine the context of undergraduate mathematics majors.

In preparing the present draft, the Task Force held several open hearings on issues concerning the undergraduate major at national meetings of the Mathematical Association of America and the American Mathematical Society. A draft of this report was mailed for review to several hundred persons including every Governor, Chair, and Secretary of the 29 Sections of the Mathematical Association of America; department heads, deans, and provosts at a variety of institutions; and many leaders of mathematical professional societies. This draft was also reviewed at meetings of AAC and MAA in January 1990. At the latter meeting, the document was discussed extensively at an invitational three-hour roundtable meeting that involved about forty experienced college and university mathematicians, including many who have been working actively to improve opportunities for women and minorities in the mathematical sciences.

Sectional Governors of the MAA were asked to nominate exemplary departments whose programs illustrate issues discussed in the report. Descriptions of some of these programs have been adapted as illustrations of promising practices in the final draft. During the spring of 1989 graduating seniors on several campuses were surveyed as part of a multi-disciplinary AAC effort to gather student opinion. Subsequently, leaders of several student chapters of the Mathematical Association of America were asked to review a draft of the report. Their careful and thoughtful responses underscore our belief in the value to students of the recommendations contained in this report.

The report has benefited enormously from these many external reviews. We believe that it now represents a consensus of the informed mathematical community concerning urgent issues of importance to the undergraduate mathematics major. In August 1990 the report was unanimously approved by the MAA Board of Governors as an official MAA statement concerning the undergraduate major. We hope that widespread discussion of this report will help focus the efforts at reform that are already underway on many campuses.

LYNN ARTHUR STEEN
Task Force Chair

Mathematics in Liberal Education

The 1985 AAC report *Integrity in the College Curriculum* [6] sets forth a vision of undergraduate education steeped in the tradition of liberal education. It describes study in depth in terms of the capacity to master complexity, to undertake independent work, and to achieve critical sophistication. To achieve the kind of depth envisioned by the authors of this report, students must grapple with connections, progress through sequential learning experiences, and enhance their capacity to discern patterns, coherence, and significance in their learning. Study in depth should enhance students' abilities to apply the approaches of their majors to a broad spectrum of problems and issues, and at the same time develop a critical perspective on inherent limitations of these approaches.

We are especially concerned in this study with how the experience in the major contributes to the education of the great majority of students who do not pursue advanced study in the field of their undergraduate major. Hence we focus more on the *quality* of students' engagement with their collegiate major than on curricular content which may be required for subsequent study or careers. This emphasis on general benefits of the major rather than on specific things learned gives the AAC undertaking a distinctive perspective that is not often emphasized in discussions of the mathematics major by mathematicians.

Historical Perspective

For over 35 years the Committee on the Undergraduate Program in Mathematics (CUPM) has helped provide coherence to the mathematics major by monitoring practice, advocating goals, and suggesting model curricula. Until the 1950s, most mathematics departments functioned primarily as service departments for science and engineering. CUPM was established in 1953 to "modernize and up-grade" the mathematics curriculum and "to halt the pessimistic retreat to remedial mathematics." At that time total enrollment in mathematics courses in the United States was approximately 800,000; each year about 4,000 students received a bachelor's degree in mathematics, and about 200 received Ph.D. degrees.

Following an unsuccessful initial effort to introduce a "universal" first-year course in college mathematics, CUPM concentrated in its second decade on proposals to strengthen undergraduate preparation for Ph.D. study in mathematics [15]. Spurred on by Sputnik and assisted by significant support from the National Science Foundation, interest in mathematics rose to levels never seen before (or since) in the United States. By 1970 total undergraduate mathematics enrollments had expanded to over three million students; U.S. mathematics departments produced 24,000 bachelors and 1,200 doctoral degrees a year.

But then the bubble burst. As student interest shifted from personal goals to financial security, and as computer science began to attract increasing numbers of students who in earlier years might have studied mathematics, the numbers of mathematics bachelor's degrees dropped by over 50% in ten years, as did the number of U.S. students who went on to a Ph.D. in mathematics. However, total undergraduate mathematics enrollments continued to climb as students shifted from studying mathematics as a major to enrolling in selected courses that provided tools necessary for other majors. In 1981, at the nadir of B.A. productivity, CUPM published its second major comprehensive report on the undergraduate curriculum [16] which advocated a broad innovative program in *mathematical sciences*.

During the 1980s the number of undergraduate majors rebounded, although both the number and the percentage of U.S. mathematics students who persist to the Ph.D. degree continued to decline [1, 21]. As a consequence, one of the most urgent problems now facing American mathematics is to restore an adequate flow in the pipeline from college study to Ph.D. attainment. Three recent reports from the National Academy of Sciences [9, 39, 44] document this crisis of renewal now confronting U.S. mathematics.

The precipitous decline in the 1970s and early 1980s in the number of undergraduate mathematics majors paralleled both a significant decrease in mathematical accomplishment of secondary school graduates and an erosion of financial support for mathematical research. Strangely, all this occurred at precisely the same time that the full scope of mathematical power was unfolding to an unprecedented degree in scientific research and business policy. These conflicting forces produced a spate of self-studies within the mathematical community, some devoted to school education [13, 19, 40, 41], some to college education [2, 50, 59], a few to graduate education [8, 12], and others to research issues [9, 24, 29]. The crisis of confidence among mathematicians reflected broad public concern about the quality and effectiveness of mathematics education [24, 32, 45]. Even research mathematicians are now taking seriously the crisis in school mathematics [27, 62]. A recent report [43] by the National Research Council recommends sweeping action based on an emerging consensus with broad support that extends well beyond the mathematical community.

Market-driven demand for undergraduate mathematics majors is strong. There is a shortage of mathematical scientists with doctor's degrees; there is consistent demand (with high salaries) for those who hold master's degrees in the mathematical sciences; there is steady demand for undergraduate mathematics majors in entry-level positions in business, finance, and technology-intensive industry; and there is increasing need for well-qualified high school teachers of mathematics, especially as a consequence of the national effort to raise standards for school mathematics. Moreover, the increasingly analytical nature of other professions (business, law, medicine) as well as the continued mathematization of science and engineering provide strong indicators of the long-range value of an undergraduate mathematics major for students who are entering other professions.

Today, mathematics is the second largest discipline in higher education. Indeed, over 10% of college and university faculty and student enrollments are in departments of mathematics. However, more than half of this enrollment is in high school-level courses, and most of the rest is devoted to elementary service courses. Less than 10% of the total post-secondary mathematics enrollment is in post-calculus courses that are part of the mathematics major. Even in these advanced courses many students are not mathematics majors—they are enrolled to learn mathematical techniques used in other fields. As a consequence, the major has suffered from neglect brought about in part by the overwhelming pressure of elementary service courses.

In spite of the 1965 and 1981 reports from CUPM and dozens of collateral studies published in the last decade, there is no national consensus about the undergraduate mathematics curriculum. However, examination of practices at many campuses reveals common threads that are highly compatible with goals of the AAC *Integrity* study. These include a multiple track system that addresses diverse student objectives, emphasis on breadth of study in the major, and requirements for depth that help students achieve critical sophistication. In this report we build on these common threads to suggest new goals for study in depth that will both enhance the mathematics major and better suit students who will live and work in the twenty-first century.

Common Goals

Mathematics shares with many disciplines a fundamental dichotomy of instructional purpose: mathematics as an object of study, and mathematics as a tool for application. These different perspectives yield two quite different paradigms for a mathematics major, both of which are reflected in today's college and university curricula. The first, reminiscent of the CUPM recommendations of the 1960s, focuses on a core curriculum of basic theory that prepares students for graduate study in mathematics. The second, reflecting the broader objectives of CUPM's 1981 mathematical sciences report, focuses on a variety of mathematical tools needed for a "life-long series of different jobs." Typically faculty interests favor the former, whereas student interests are usually inclined towards the latter. Most campuses support a mixed model representing a locally devised compromise between the two standards, although accurate descriptions of present practice are virtually non-existent. (Partly in response to the paucity of information about degree programs in mathematics, the National Research Council through its project Mathematical Sciences in the Year 2000 is seeking to stimulate a comprehensive examination of degree patterns in mathematics.)

The 1985 AAC *Integrity* study, echoing themes reflected in similar studies [10, 38, 42], provides a broad context in which to examine practices of individual disciplines. The AAC report highlights "study in depth" as an essential component of liberal education, but not for the reasons commonly advanced by students and their parents—preparation for vocation, for professional study, or for graduate school. *Integrity* views study in depth as a means to master complexity, to grasp coherence, and to explore subtlety. "Depth cannot be reached merely by cumulative exposure to more and more ...subject matter." The AAC goals for study in depth are framed by twin concerns for intellectual coherence intrinsic to the discipline and for development of students' capacity to make connections, both within their major and with other fields.

Both previous models advanced by CUPM for the undergraduate major reflect these same potentially conflicting concerns. The earlier major, which emphasized a traditional core mathematics curriculum as preparation for graduate study, was motivated principally by the internal coherence of mathematics; the more recent proposal stressed an interplay of problem solving and theory across both the broad spectrum of the mathematical sciences and the boundaries that mathematics shares with other disciplines. Although many believe that the latter model, emphasizing broadly applicable mathematical methods, is better geared to graduates' future employment, others—including many liberal educators—believe that broader abilities such as the art of reasoning and the disposition for questioning are of greater long-term practical value. However, both types of majors—and the many mixtures that exist on today's campuses—help students discern patterns, formulate and solve problems, and cope with complexity. In this sense, present practice of mathematical science majors in U.S. colleges and universities matches well the overall objectives of *Integrity in the College Curriculum*.

Certain principles articulated by CUPM in 1981 (reprinted in a recent compendium [17] of CUPM curriculum recommendations from the past decade) make explicit areas where *Integrity*'s objectives and those of the mathematical community align:

- The primary goal of a mathematical sciences major should be to develop a student's capacity to undertake intellectually demanding mathematical reasoning.

- A mathematical sciences curriculum should be designed for all students with an interest in mathematics, with both appropriate opportunities for average mathematics majors and appropriate challenges for more advanced students.
- Every student who majors in the mathematical sciences should complete a year-long course sequence at the upper-division level that builds on two years of lower-division mathematics.
- Instructional strategies should encourage students to develop new ideas and discover new mathematics for themselves, rather than merely master the results of concise, polished theories.
- Every topic in every course should be well motivated, most often through an interplay of applications, problem-solving, and theory. Applications and interconnections should motivate theory so that theory is seen by students as useful and enlightening.
- Students majoring in mathematics should undertake some real-world mathematical modelling project.
- Mathematics majors should complete a minor in a discipline that makes significant use of mathematics.

Emphasis on coherence, connections, and the intellectual development of all students are evident in these principles from the 1981 CUPM report. At the level of broad goals, the prevailing professional wisdom concerning undergraduate mathematics matches well the intent of AAC's *Integrity*.

Diverse Objectives

Once one moves beyond generalities and into specifics of program development, however, mainstream mathematical practice often diverges from many of the explicit AAC goals. Most students study mathematics in depth not to achieve broad goals of liberal education but for some professional purpose—for example, to support their study of science or to become a systems analyst, teacher, statistician, or computer scientist. Others study mathematics as a liberal art, an enjoyable and challenging major that can serve many ends. It is as true in mathematics as in any other field that the great majority of undergraduate mathematics majors do not pursue advanced study that builds on their major.

Because so few U.S. students pursue graduate study in the mathematical sciences, many mathematicians believe that the mathematics major should be strengthened in ways that will prepare students better for graduate study in mathematics. Whereas formerly this view may have been based on the myopic academic view of undergraduate education as the first step in the reproduction of university professors, today in mathematics this perspective is reinforced by vigorous argument based on an impending shortage of adequately trained mathematics faculty. Some argue for greater depth to ensure that all mathematics majors are capable of pursuing further study; others argue for greater breadth to attract more students to the mathematical sciences. National need has now reinforced self-interest in emphasizing preparation of the next generation of Ph.D. mathematicians as an important priority for many college and university mathematics departments [20, 39].

This said, there remains considerable room for debate about strategies for achieving the several different (but overlapping) objectives that are common among the majors offered by the 2500 mathematics departments in U.S. colleges and universities:

- ADVANCED STUDY. Preparation for graduate study in various mathematical sciences (including statistics and operations research) or in other mathematically based sciences (e.g., physics, computer science, economics).
- PROFESSIONAL PREPARATION. Skills required to pursue a career that requires considerable background in mathematics:
  - *Natural and Social Science.* Background for careers in science or engineering, in biology (including agriculture, medicine, and biotechnology), in many computing fields, and in such emerging interdisciplinary fields as cognitive science or artificial intelligence.
  - *Business and Industry.* Preparation for careers in management, finance, and other business areas that use quantitative, logical, or computer skills normally developed through undergraduate study of mathematics.
  - *School Teaching.* Preparation for teaching secondary school mathematics, or for careers as mathematics-science specialists for elementary school.
- LIBERAL EDUCATION. General background for non-mathematical professions such as law, medicine, theology, or public service, or for other employment which does not directly use mathematical skills.

There is no lack of sources for advice about objectives. Recent documents outline priorities for teacher training [14], applied mathematics [25], and service courses [33]. The diversity of U.S. education ensures that different departments will have different priorities that recognize differences in entering students, in goals of graduates, and in institutional missions.

For many students the link between their undergraduate major and their post-graduate plans is very elastic. Such students study mathematics for the same reason that hikers climb mountains—for challenge, for fun, and for a sense of accomplishment. Often mathematics is paired with another major to provide complementary strengths. At one institution this has helped strengthen the mathematics major while enhancing connections to other fields of study:

> An analysis of the senior class shows that nearly half of the graduates had double majors. In many cases the mathematics major was taken to provide theoretical and computational grounding necessary for a modern approach to primary majors in economics, physics, chemistry, or biology. In other cases availability of a double major gave students who enjoyed mathematics an opportunity to continue their studies in mathematics while also gaining a desired major in an apparently unrelated field—for example, English, philosophy, or music. Many of our majors chose mathematics as a sound liberal arts approach to a general education. The success of the major program in mathematics is due in part to the belief of both faculty and students that the study of mathematics is not just for those intending to pursue a career in the area.

Some argue that the goals of liberal education are best served by a mathematics major designed to prepare students for graduate school. Even though most mathematics majors never undertake further study of mathematics, advocates of pre-graduate preparation argue that the special combination of robust problem solving with rigorous logical thinking achieved by a solid pre-graduate major also serves well the more general objectives of sequential study, intellectual development, and connected learning. This view is substantiated in part by a strong history of mathematics majors being eagerly sought by employers and graduate programs in other areas (e.g., law, economics) for a wide variety of non-mathematical professions.

Others argue that since most of today's college students do not foresee graduate study in mathematics as a desirable goal, it is only through a more general major stressing links to other fields that enough students can be recruited to major in mathematics to serve as a proper basis for the nation's needs in teaching, in science, and in mathematics itself. In this view, a broad major stressing mathematics as part of liberal education is an effective strategy for strengthening the pool of potential mathematical doctoral students as well as students in other mathematically oriented professions. It is rare to find a department of mathematics that would naturally place top priority on a major that specifically serves liberal education, as is common, for example, in many departments of English or philosophy. The needs of society and the constraints of client disciplines in science and engineering do not permit mathematicians this luxury. But in the present climate, departments are rediscovering the strategic value of a broad major, even for those who do continue professionally in mathematics. Here's how one institution expresses its objectives for the mathematics major:

> The mathematics major is designed to include students with a wide variety of goals, tastes, and backgrounds. Mathematics is an excellent preparation for fields from technical to legal, from scientific to managerial, and from computational to philosophic. It is also a source of satisfaction for people in every line of endeavor. Recognizing this, we have constructed a program to welcome interested students of all sorts. Our comparatively unstructured program reflects not only the diversity of interests of our students, but their increasingly diverse backgrounds, and the increasingly diverse nature of what are now being called the "mathematical sciences."

Since department goals must match institutional missions, it would not be right for any committee to recommend uniform goals for individual departments. We can, however, urge increased attention to an important and distinctive feature of undergraduate mathematics: *National need requires greater encouragement for students to continue their study of mathematics beyond the bachelor's level—whether as preparation for school teaching, for university careers, or for employment in government and industry.* In some institutions this encouragement may arise from a thriving pre-graduate program; in others it may evolve from an emphasis on liberal education. In all cases, departmental objectives must be realistically matched to student aspirations and to institutional goals. Wherever faculty and students share common objectives, mathematics can thrive.

Multiple Tracks

Most mathematics departments resolve the dilemma of diverse goals for the major with some sort of track system. In some institutions there are separate departments such as Applied Mathematics, Operations Research, Statistics, or Computer Science, whereas in others these options are accommodated by means of explicit or implicit choices within the offerings of a single Department of Mathematics or Department of Mathematical Sciences. Tracks within the major are a sensible strategy to respond to competing interests of students, of faculty, and of institutions.

Although tracks do accommodate student interests—and thereby help sustain enrollments—they can produce a fragmented curriculum. Whereas in the late 1960s most mathematics majors took six or seven standard courses before branching into electives, by 1981 CUPM found that there was no longer any national agreement on such an extensive core of the undergraduate mathematics major. At that time, only elementary calculus and linear algebra were universally recognized as required courses within the mathematics major. Branching occurred after the third or fourth course, rather than after the sixth or seventh.

Today however, despite institutional diversity, there is striking uniformity in the elementary and intermediate courses pursued by mathematics majors: all begin with calculus for two, three, or four semesters; most introduce linear algebra in the sophomore year and require one or two semesters of abstract algebra; virtually all require some upper division work in analysis—the "theory of calculus." Nowadays, most require some computer work (programming, computer use, or computer science) as well as some applied work (statistics, differential equations, etc.) among the electives. This restores a *de facto* six-course core to the major, typically half the total.

The rise, fall, and restoration of a core curriculum in the mathematics major paralleled similar patterns in other arts and sciences. Whereas the CUPM recommendations of the late 1960s may have had too narrow a focus, the subsequent curricular chaos of the mid-1970s may have been too unstructured. Mathematicians began worrying then—as AAC is now—about whether the typical student's experience with the mathematics major may lack appropriate coherence and depth. Too often, it seemed to many, the mathematics major had become just an accumulation of courses without sufficient structure to ensure a common core of learning.

In most colleges courses in the mathematical sciences are taught in many different departments. Even upper division mathematics courses are commonly taught in departments of application: a recent survey [28] shows that as many students study post-calculus mathematics outside departments of mathematics as do so within traditional mathematics course offerings. Examples include discrete mathematics taught in departments of computer science; methods of mathematical physics taught in departments of physics; logic and model theory taught in departments of philosophy; optimization and operations research taught in departments of economics; and mathematical foundations of linguistics taught in departments of linguistics. The practice of cross-listing such courses or counting them as electives in the major varies enormously (and arbitrarily) from campus to campus.

Whether for good or ill, the diffusion of mathematics courses both within departments of mathematical sciences and into other departments has moved the mathematics major away from a strict linear vertical pattern towards a more horizontal structure typical of the humanities or social science major. Today's major, however, retains a distinctive strength of mathematics: sequenced learning. By its very nature, mathematics builds on itself and reinforces links among related fields. All mathematics courses build on appropriate prerequisites. A student who progresses from calculus to probability to operations research sees just as many connections as does one who moves through the more traditional sequence of advanced calculus and real analysis. Although the focus of each student's work is different, the contributions made by each track to the general objectives of study in depth are comparable, and equally valuable.

Moreover, it is common for advanced courses to be offered in sequences (e.g., Abstract Algebra I, II; Real Analysis I, II; Probability and Statistics I, II) that begin with a three or four course chain of prerequisites. Many departments, following the 1981 CUPM recommendations, require mathematics majors to take some advanced sequence without specifying which particular sequence it should be. Thus most mathematics majors today take a substantial sequence of courses, but they no longer all take the *same* sequence of core courses. This is a wise policy for undergraduate mathematics in today's diverse climate: *Each student who majors in mathematics should experience the power of deep mathematics by taking some upper-division course sequence that builds on lower-division prerequisites. It is neither necessary nor wise, however, to require that all mathematics majors take precisely the same sequence.*

Flexibility with rigor can be administered in a variety of ways. In one institution with a flourishing mathematics major, the mechanism is a personal "contract" developed to suit each student's own objectives:

> Students arrange their major sequence according to a contract system. Potential majors meet with a member of the department and prepare a list of courses and activities that will constitute the major. This allows the student to arrange his or her program to suit special needs. The faculty member judges the appropriateness of the student proposal in terms of post-graduate plans, other studies, and general departmental guidelines. This contract system has two distinct advantages: it serves the personal needs of students, and the process itself enhances students' education. The process of developing the contract provides an opportunity for the student to work closely with a faculty member, to understand the variety of mathematical options in a personal framework, and to see how a program ensuring depth and breadth of study can be achieved.

Emphasizing Breadth

At the same time as we stress the value of sequential courses to study in depth, we must also emphasize the essential contribution of breadth to building mathematical insight and maturity. Whereas course sequences demonstrate depth by building in expected fashion on prior experience, the links that emerge among very different courses (tying geometry to calculus, group theory to computer science, number theory to analysis) reveal depth by indirection: such links point to deeper common principles that lie beyond the student's present understanding but are within grasp with further study. They show the mountain yet to be climbed—to shift metaphors from depth to height—and offer hints of the explanatory panorama to be revealed by some future and more profound principles.

There are other good reasons to recommend breadth as an important objective of an undergraduate mathematics major. Students who are introduced to a variety of areas will more readily discern the power of connected ideas in mathematics: unexpected links discovered in different areas provide more convincing examples of a deep logical unity than do the expected relationships in tightly sequenced courses.

For the many majors who will teach (either in high school or college), it is vitally important that their undergraduate experience provide a broad view of the discipline—since further study generally is more narrow and specialized. For those seeking their niche in the world of mathematics, a broad introduction to many different yet interconnected subjects, styles, and techniques helps pique interest and attract majors. And for the many students who may never make professional use of mathematics, depth through breadth offers a strong base for appreciating the true power and scope of the mathematical sciences. Graduates of programs that emphasize breadth will become effective ambassadors for mathematics.

Every student who majors in mathematics should study a broad variety of advanced courses in order to comprehend both the breadth of the mathematical sciences and the powerful explanatory value of deep principles. Such breadth can sometimes be achieved with courses offered by the department of mathematics, but more often than not it would be educationally advantageous for students to also select a few mathematically-based courses offered by other departments.

Effective Programs

Departments of mathematics in colleges and universities exhibit enormous variety in goals and effectiveness. In various universities, the percentage of bachelor's degrees awarded to students with majors in mathematics ranges from well under one-half of 1 percent to over 20 percent. In some departments the major is designed primarily to prepare students for graduate school. Other departments focus much of their major on preparing students to teach high school mathematics, or on preparing students for employment in business and industry. Most departments fail to attract or retain many Afro-American, Hispanic, or Native American students, whereas a few succeed in this very difficult arena.

Many measures can be used to monitor effectiveness of a mathematics major. Indicators of numbers of majors, of employability, of graduate school admissions, of eventual Ph.D.s, or of placement in teaching jobs are used by different departments according to their self-determined missions. Many mathematics departments work hard to improve their effectiveness in one or another of these different dimensions. Exploration, experimentation, and innovation—along with occasional failures—are the hallmarks of a department that is committed to effective education.

Mathematics programs that work can be found in all strata of higher education, from small private colleges to large state universities, from average to highly selective campuses. The variety of mathematics programs that work reveal what can be achieved when circumstance and commitment permit it. When faculty resolve is backed by strong administrative support, most mathematics departments can easily adopt strategies to build vigorous majors even while meeting other service obligations.

One department that has had great success in attracting students to major in mathematics bases its work on two "articles of faith:"

- We believe that faculty should relate to their students in such a way that each student in the department will know that someone is personally interested in him and his work.
- We believe that careful and sensitive teaching that helps students develop confidence and self-esteem is far more important than curriculum or teaching technique.

Another department builds strength on a foundation of excellent introductory instruction:

> We put our best teachers in the introductory courses. We put the most interesting material in the introductory courses. We try to make the statements of problems fun, not dry. We work very hard to motivate all topics, drawing on applications in other disciplines and in the working world. We are less interested in providing answers than in motivating students to ask the right questions.

Effective mathematics programs reflect sound principles of psychology as much as important topics in mathematics:

> We try to make students proud of their efforts in mathematical problem-solving, and especially proud of their partial solutions—what some might call mistakes. We look at how much is right in an answer and teach how to detect and correct the parts that are wrong.

Regular, formal recognition of student achievement at different stages of the major serves to build students' confidence and helps attract students to major in mathematics. Students know mathematics' reputation for being challenging, so recognition of honest accomplishment can provide a tremendous boost to a student's fragile self-esteem. *Effective programs teach students, not just mathematics.*

Challenges for the 1990s

Changes in the practice of mathematics and in the context of learning pose immense challenges for college mathematics. Many of those issues that pertain directly to course content, curricular requirements, and styles of instruction are under review by committees of the mathematical community. We focus here on challenges that transcend particular details of courses and curriculum:

- The *learning* problem: To help students learn to learn mathematics.
- The *teaching* problem: To adopt more effective styles of instruction.
- The *technology* problem: To enhance mathematics courses with modern computer methods.
- The *foundation* problem: To provide intellectually stimulating introductory courses.
- The *connections* problem: To help students connect areas of mathematics and areas of application.
- The *variety* problem: To offer students a sufficient variety of approaches to a mathematics major to match the enormous variety of student career goals.
- The *self-esteem* problem: To help build students' confidence in their mathematical abilities.
- The *access* problem: To encourage women and minorities to pursue advanced mathematical study.
- The *communication* problem: To help students learn to read, write, listen, and speak mathematically.
- The *transition* problem: To aid students in making smooth transitions between major stages in mathematics education.
- The *research* problem: To define and encourage appropriate opportunities for undergraduate research and independent projects.
- The *context* problem: To ensure student attention to historical and contemporary contexts in which mathematics is practiced.
- The *social support* problem: To enhance students' personal motivation and enthusiasm for studying mathematics.

These challenges have more to do with the success of a mathematics program than any curricular structure. In the diverse landscape of American higher education, successful programs differ enormously in curricular detail, but they all have in common effective responses to many of these broader challenges. The agenda for undergraduate mathematics in the 1990s must focus at least as much on these issues of context, attitude, and methodology as on traditional themes such as curricula, syllabi, and content.

Learning

One principal goal of the undergraduate mathematical experience is to prepare students for life-long learning in a sequence of jobs that will require new mathematical skills. Departments of mathematics often interpret that goal as calling for breadth of study. But another interpretation is just as important: because mathematics changes so rapidly, undergraduates must become independent learners of mathematics, able to continue their own mathematical education once they graduate.

Most college students don't know how to learn mathematics, and most college faculty don't know how students do learn mathematics. It is a tribute to the efforts of individual students and teachers that any learning takes place at all. Effective programs pay as much attention to learning as they do to teaching.

First-year students need special attention to launch their college career on a suitable course. Typically, they carry with them a high school tradition of passive learning which emphasizes bite-sized problems to be solved by techniques provided by the textbook section in which the problem appears. Unfortunately, by maintaining this traditional teaching format which perpetuates the myth of passive mathematics learners, college calculus teachers typically contribute more to the problem than to its solution.

For example, calculus, the common entry point for potential mathematics and science majors, often fails to come alive intellectually as it should or as it is now at many institutions where calculus reform efforts are underway. One school has found that new goals for calculus can significantly enhance the entrée of students into the study of college mathematics:

> The larger goals of the major are reflected in the calculus sequence, which is founded on three principles: context, collaboration, and communication. "Context" means that we focus on the meaning and significance of calculus in the world. "Collaboration" means that students work in groups and support each other. "Communication" means the recognition that calculus is first of all a language, not only for scientists, but for economists and social scientists. Our goal is fluency.

Another institution uses calculus as a vehicle to broaden radically the view of mathematics that students bring with them from high school:

> Calculus should give students a solid base for advanced study. It is our opinion that our calculus courses were the weakest part of our program. We had, in effect, allowed the high schools to set the tone for our entire program. Our new course is so radically different from traditional calculus that our students are forced to confront the transition from school to college mathematics. It carries several important messages, e.g., mathematics is crucial for understanding science; mathematics has a strongly experimental side; mathematics is something we all are capable of understanding deeply; and mathematics is the most powerful of all the sciences.

Some institutions offer special freshman seminars as a way to encapsulate the ideal of liberal education in an intimate setting that permits students to identify with faculty mentors. However, in mathematics the massive tradition of calculus often stands in the way, so very few mathematics majors can trace the origin of their college major to a freshman seminar. Ideally calculus itself would be seen by colleges as the intellectual equivalent of a freshman seminar in which students learn to speak a new language. If that analogy were to be accepted, mathematics departments would teach calculus only in a context that placed a great deal of emphasis on one-to-one communication between student and teacher. Unfortunately, in too many institutions calculus is taught in large impersonal settings that make meaningful person-to-person dialogue unrealistic. Many efforts are now underway to reform the teaching of calculus [64]; most of these experiments emphasize student motivation and styles of learning as a primary factor in reshaping the course.

One way or another, students should learn early in their college years how to study and learn mathematics. They should learn psychological as well as mathematical strategies for solving problems. They should come to recognize that it is common even for mathematicians to hear lectures or read material that they cannot grasp, and they must learn how to pick up clues from such experiences that will fit into their personal mathematical puzzles only some time later. They should learn the value of persistence and the strategic value of going away and coming back. These "metacognitive" skills to control one's own learning are virtually never learned in high school mathematics, so they must be planned into the early stages of the college curriculum.

As students progress through their mathematical study, they need to learn the value of library and electronic resources as tools for learning. Mathematics students rarely use the library or other sources of information, concentrating instead on mastering material in course texts. They need specific assignments that focus on the big map of mathematics in order to gain perspective on their brief undergraduate tour. *Undergraduate students should not only learn the subject of mathematics, but also learn how to learn mathematics.* The major in mathematics should become more than the sum of its courses. By conscious effort to help students negotiate in unfamiliar terrain, instructors can provide them with the tools of inquiry necessary to approach the literature and learn whatever they need to know.

Teaching

The purpose of teaching, and its ultimate measure, is student learning. So in some sense one cannot discuss one without the other. However, as students must learn to learn, so teachers must learn to teach. In mathematics more than in most other subjects, the role of teaching assistants and part-time instructors is particularly important, especially in the first year [12]. Although there is no formula for successful teaching, there is considerable evidence that separates certain practices that have proven successful from those that are generally ineffective [55]. Teachers who study this evidence can learn much from the experiences of others.

Despite the general reputation of mathematics as one of the most desirable environments for developing rigorous habits of mind, criticism of undergraduate mathematics has been mounting in recent years for failure in this, mathematics' distinctive area of strength. Those who study cognitive development criticize standard teaching practices for failing to develop fully students' power to apply their mathematical knowledge in unfamiliar terrain. These critics conclude that present teaching practice in undergraduate mathematics does not do as much as it should to develop students' intellectual power.

The evidence of failure is persuasive, both locally and globally [57]. Data on the inability of the profession to attract and retain the best and brightest college graduates is confirmed by case studies of students who cannot make effective use of what they have learned. Although some very good students use a mathematics major as a platform for substantial accomplishment, the majority of those who major in mathematics never move much beyond technical skills with standard textbook problems. Passive teaching and passive learning results from an unconscious conspiracy of minimal expectations among students and faculty, both of whom find advantages in a system that avoids the challenges of active learning that fully engages both students and teacher. Both the curriculum and teaching practices must respond to this challenge of intellectual malnutrition that is all too common in today's major.

Much of the research that bears on how students learn college mathematics has been conducted in the setting either of high school mathematics or college physics. The results from these efforts are often surprising, yet not well known among university mathematicians. They show, among other things, that formal learning by itself rarely influences behavior outside the artificial classroom context in which the concept was learned [53, 54]. Students who know how to solve differential equations of motion often have no better insight into the behavior of physical phenomena described by these equations than do others who never studied the equations; students who have learned course-based tests of statistical significance frequently do not recognize statistical explanations for events in the world around them [47].

Additional evidence of how young adults learn mathematics—or more often, why they fail to learn—has accumulated in recent years as a result of many innovations in teaching tried on different campuses. For example, intervention programs designed to improve the mathematical performance of minority students show the importance of a supportive environment: constructive teamwork in a context of challenging problems in which instructors and students know each other personally builds mathematical self-esteem and, as a consequence, leads to greatly improved learning [5, 30, 63]. Very different but equally striking lessons emerge from experiences of students who study calculus in a technology-intensive environment: by forcing students (and instructors) to focus on the behavior of mathematical objects (functions, algorithms, operators) rather than on their formalism, and by integrating visual, numerical, and symbolic clues into the mathematical environment, computers reveal to students and faculty both avenues for insight and common sources of misconception [36].

A third example, but by no means the last that could be cited, can be found in evidence of improved student motivation and self-reliance that occurs in those contexts where research-like experiences are used to enrich traditional classroom and textbook experiences: students whose minds and eyes become engaged in the challenge of true discovery are frequently transformed by the experience [56].

The evidence from such diverse but non-traditional instructional environments shows clearly the effectiveness of instruction that builds self-confidence on the foundation of significant accomplishment in a context that is meaningful to the student. Here is an especially dramatic example:

> In 1986 we began a critical evaluation of our program, course offerings, and teaching methods. This examination led to profound changes in our understanding of the teacher-student relationship, and of our role in the educational process. We found, for example, that we had not engaged our students sufficiently to assume an active role in their learning of mathematics. So we deliberately modified our courses and attitude to experiment with active student participation in doing mathematics problems and theory both in class and outside class.

> Results were strikingly positive, and we largely discontinued the typical mathematics lecture format, since lecturing kept students in a passive role. With an active participation method, students studied the text and worked problems before class; faculty and students discussed difficult points in class. Students presented problems and results on the board in class with encouragement and guidance from the instructor. We found that students became actively and enthusiastically involved in their learning of mathematics, with the instructor acting as a coach.

> As a consequence of these changes, our faculty and students have become a community of learners and scholars. Students now do mathematics in groups outside class, and more graduating seniors are seeking advanced degrees in mathematics. The number of mathematics majors rose from 69 in 1986 to 104 in 1988; the Mathematics Department is now the largest unit in the School of Natural Sciences. Finally, and perhaps most important, faculty affirm the belief that many more students can realize their mathematical abilities.

Several barriers separate educational studies and experiments from the larger community of college and university mathematicians. First, there are very few individuals who conduct formal research dealing directly with college mathematics. Second, mathematicians tend to distrust educational research. Third, and perhaps more important, mathematicians follow habit more than evidence in their teaching styles: even well-documented reports of better methods are insufficient to influence mathematicians to change their teaching habits. (This is not really too surprising, since neither do convincing classroom explanations of effective mathematical methods suffice to eradicate deep-seated misconceptions among students.)

Too often mathematicians assume with little reflection that what was good for their education is good enough for their students, not realizing that most of their students, not being inclined to become mathematicians, have very different styles of learning. College faculty must begin to recognize the proven value of various styles of instruction that engage students more directly in their own learning. *Those who teach college mathematics must seek ways to incorporate into their own teaching styles the findings of research on teaching and learning.*

Studies of metacognition and problem solving have yielded some insights that could be useful in pedagogy, but they have also been frustrated by barriers that confront all teachers of mathematics (for example, the difficulty of assessing just what has been learned, and the great length of time required to develop effective problem-solving heuristics). Such studies may yield insights that will change for the better the way teachers teach and students learn. But so far, college-level evidence is sufficiently slim to make the case unconvincing to those who most need to be persuaded. We really don't know how to induce most students to rise to the challenge of mathematical thinking; we have much to learn about what works and what does not. *To improve our understanding of the intellectual development of college mathematics students, mathematicians should increase their efforts to conduct research on how college students learn mathematics.*

We need to experiment with new ways to evaluate teaching. One key factor in good teaching is how much students learn; other factors include such issues as how many students decide to major in mathematics, to go on to graduate school, or to work in mathematical careers. These are measures of the quality of teaching done both by an individual and by a department. They look not only to indicators such as demonstrable knowledge, but also to motivation, attitude, and enthusiasm for the discipline. *Evaluation of teaching must involve robust indicators that reflect the broad purposes of mathematics education.*

Technology

Computing has changed profoundly—and permanently—the practice of mathematics at every level of use. College mathematics departments, however, often lag behind other sciences in adapting their curricula to computing, although considerable momentum is now building within the community for greater use of computing. The delay in response may have been due in part to conservatism of mathematicians, but at least as important is the simple fact of computer power: only in the last few years have desk-top machines achieved sufficient power to provide a legitimate aid to undergraduate (and research) mathematics. As a consequence, scientific computation is becoming a third paradigm of scientific investigation—alongside experimental and theoretical science—and the role of experiment in the practice of mathematics itself is increasing [52].

Computing can enhance undergraduate study in many ways. It provides natural motivation for many students, and helps link the study of mathematics to study in other fields. It offers a tool with which mathematics influences the modern world and a means of putting mathematical ideas into action. It alters the priorities of courses, rendering certain favorite topics obsolete and making others, formerly inaccessible, now feasible and necessary [34, 68]. Computers facilitate earlier introduction of more sophisticated models, thus making instruction both more interesting and more realistic. The penetration of computing into undergraduate mathematics is probably the only force with sufficient power to overcome the rigidity of standardized textbooks [22, 59, 66].

The power of technology serves also an epistemological function by forcing mathematicians to ask anew what it means to know mathematics. Those who explore the impact of technology on education indict introductory mathematics courses for imparting to students mostly skills that machines can now do more accurately and more efficiently. It is certainly true that typical indicators of student performance document primarily that mathematics students can carry out prescribed algorithms—just what computers (or calculators) can do. College faculty can no longer avoid the deep challenge posed by computers for undergraduate mathematics: once calculations are automated, what is left that can be taught effectively to average students?

Responses to this challenge are taking shape in experimental programs in many departments of mathematics. It is, therefore, too early to describe the impact computing will have on the mathematics major. Certainly in those courses and tracks devoted to applied mathematics, computing must exert a major influence on the shape of the curriculum. In this age it would be unconscionable to offer a major in applied mathematics, statistics, or operations research without substantial and fully integrated use of computer methods. Change will come more slowly in core subjects such as topology, analysis, and algebra. In each of these subjects there are impressive computer-based applications (e.g., fractals, coding theory, dynamical systems), yet none of these applications has been central to the traditional methodology of the subject as taught in introductory courses. Despite differences in the pace of change, however, there is no turning back: computers have dramatically altered the practice of mathematics. *To ensure an effective curriculum for the twenty-first century, undergraduate mathematics should change—both in objectives and in pedagogy—to reflect the impact of computers on the practice of mathematics.*

Early experiments that make significant use of computing in undergraduate mathematics courses show that as the balance of student work shifts from computation—which machines do better than humans—to thought, the course becomes more difficult, more unsettling, and less closely attuned to student expectations [58]. As the ground rules of mathematics change from carrying out prescribed procedures to formulating problems and interpreting results, it will become more important than ever for faculty to communicate clearly to students the goals of the curriculum and how they might differ from what students have been led to believe by their prior study of school mathematics.

One institution reports that computers have changed the context of education in significant and unexpected dimensions:

> We constructed a strong computer-experimental component at all levels. Besides the obvious advantages for building experience, context, and intuition, there are less obvious payoffs. For example, laboratories are places where students spend lots of time and which become, in reality, their habitats outside of their dormitory rooms. Students form allegiances and friendships in laboratories.

Different types of surprises were revealed on another campus that has pioneered use of computers in advanced courses:

The use of computer software made possible the introduction of topics previously reserved for graduate students. Examples include the use of MACSYMA, REDUCE, and MACAULEY in commutative algebra and algebraic geometry. For example, a 1989 honors thesis gave us strong evidence of the advances possible in learning mathematics with the help of computational aids.

Computers change not only how mathematics is practiced, but also how mathematicians think. Both changes are unsettling, yet ripe with opportunity for effective education. Indeed, in the realm of computing, students and faculty must grope together towards a new balance of power among the many components of undergraduate mathematics.

The transition of mathematics from a purely cerebral paper-and-pencil (or chalk-and-blackboard) discipline to a high technology laboratory science is not inexpensive. Space must be expanded for laboratories; classrooms and offices need to be equipped with computers and display devices; support staff must be hired to maintain both hardware and software; faculty must be given time to learn to use computers, to learn to teach with computers, and to redesign courses and entire curricula to reflect the impact of computing. Institutions must plan not only for an expensive transition, but also for continued operation at a higher plateau comparable to the traditional laboratory sciences. *Colleges must recognize in budgets, staffing, and space the fact that undergraduate mathematics is rapidly becoming a laboratory discipline.*

Foundations

Because of the highly sequenced nature of the mathematics curriculum, no student can complete an undergraduate mathematics major without having secured a proper foundation of calculus, linear algebra, and computing in the first two years of college. For many students, half of the credits required for the major are taken in the first two years. So the nature of mathematical learning in these years is of crucial importance both for individual success in completing a strong mathematics major, and for programmatic success in building a critical mass of upper-class mathematics majors.

One-third of the first and second year college students in the United States are enrolled in two-year colleges, including over two-thirds of Afro-American, Hispanic, and Native American students. It is clear from these figures that any effort to strengthen the undergraduate mathematics major, especially to recruit more majors among minority students, must be carried out in a manner that includes two-year colleges as a full partner in preparing the foundation for study in depth.

The tradition of common texts and relatively standard syllabi for standard mathematics courses in the first two years has facilitated transfer of students and credits during these years even as it has mitigated against the open intellectual environment many believe to be essential for effective learning. Now, however, as momentum builds for reform of courses in the first two years, and as departments experiment in an effort to reshape the entire mathematics major, there is some risk that students from lower socio-economic backgrounds—the predominant clientele of the two-year colleges—will find themselves pursuing a course of study that is inconsistent with the efforts of four-year colleges to improve the undergraduate mathematics major.

Some four-year institutions that are engaged in curricular reform are extending the scope of their mathematics program to include informal consortia with other nearby institutions.

One private Eastern liberal arts college is building just such arrangements into its mathematics program:

> Plans are underway to create a partnership with a local community college and a public school system to interest students, especially minority students, in mathematics.

Many institutions maintain regular ties with local high schools or community colleges, but it is rare to find such arrangements related specifically to mathematics departments. What is now rare should become common: *To ensure equal opportunity for access to undergraduate mathematics majors, mathematics departments should work with nearby two-year colleges to maintain close articulation of programs.*

Connections

Recent studies of the mathematical sciences [7, 8] point to two special features that have characterized twentieth-century research: the extensive growth in areas of application (no longer just limited to physics and engineering) and the impressive unity of mathematical theories (revealed by the frequent use of methods from one specialty to solve problems in another specialty). Connectedness, therefore, is inherent in mathematics. It is what gives mathematics its power, what establishes its truth, and what reveals its beauty.

Mathematics is widely recognized as the language of science. Its enabling role in the development of the physical sciences formed the paradigm of the scientific method. Today it is beginning to play a similar role in the biological sciences, where mathematical tools as diverse as knot theory, nonlinear dynamics, and mathematical logic are being applied to model the structure of DNA, the flow of blood, and the organization of the brain.

Similar connections have emerged in the human, social, and decision sciences. Statistical models undergird virtually every study of human behavior; axiomatic studies have helped establish a rigorous theory of social choice; and multi-dimensional mathematical analysis is employed widely to model the multitudinous attributes of economic, psychological, or social behavior. Today mathematics is truly the language of all science—physical, biological, social, behavioral, and economic.

Even as the connections multiply between abstract ideas of mathematics and concrete embodiments in the world, so too have the internal connections within the mathematical sciences proliferated. Key theorems and deep problems that link separate mathematical specialties have provided a force for vast growth of interdisciplinary research. Examples abound, including such areas as stochastic differential equations at the interface of probability theory and analysis; combinatorial geometry joining arithmetical methods of discrete mathematics to problems of space, shape, and position; and control theory that employs tools from analysis, linear algebra, statistics, and computer science to formulate effective mechanisms of control for automated processes.

If the undergraduate major does not reveal connections, it has not revealed mathematics. Most mathematics courses and most mathematics majors do make substantial contributions to this objective. Indeed, it is not uncommon for sophomores to select mathematics as a major instead of chemistry or biology precisely because in their mathematics courses they can see more clearly the logical connections among different parts: in mathematics they can "figure things out" rather than just memorizing results. (Of course, many students make the opposite choice, but usually for other reasons.)

At its best, mathematics overflows with connections, both internal and external. But one must be honest: undergraduate courses do not always show mathematics at its best. At their worst, especially in lower-division courses through which both majors and non-majors must pass, they reveal mathematics as a bag of isolated tricks: problems in elementary courses are often solved more by recognition of which section of the text they come from than by any real understanding of fundamental principles. *Dealing with open-ended problem situations should be one of the highest priorities of undergraduate mathematics.* For example:

- Mathematics teachers could bring in outside ("real-world") examples to illustrate applications of material being studied in regular coursework.
- Student projects could emphasize connections, either to fields that use mathematics or from one part of mathematics to another.
- Greater emphasis on multi-step problems amenable to a variety of approaches would wean students away from the school tradition of bite-sized, self-contained problems.
- Problem-oriented seminars provide wonderful opportunities to explore links between various branches of mathematics.

Such problems would be pregnant with ambiguity, ripe with subtle connections, and overflowing with opportunities for multi-faceted analyses.

Variety

Mathematicians are fond of talking about an elusive concept called "mathematical maturity" that is the Holy Grail of undergraduate mathematics [60]. Maturity is one objective of study in depth, but its meaning must be derived from the context of a student's level and goals. Depth itself is a metaphor for many things. To a mathematician it signals knowledge, insight, complexity, abstraction, and proficiency; to some others it connotes such elusive concepts as ownership, empowerment, and control. Although most colleges equate study in depth with the major—a circumstance reflected also in this report—it is important to recognize that for some students the major may not achieve the objectives that many have for study in depth. For these students, curricular structures other than the traditional major may better approach their goals for study in depth.

Many college students study mathematics as an important adjunct to another field which is their primary interest (e.g., economics, education, biology). Some colleges offer joint majors that combine study of mathematics with study in a related field, usually tied together with some type of joint project. The ever-present danger in such options is that they merely combine two shallow minors without ever achieving the depth traditionally required in a major. Notwithstanding this risk, one must acknowledge that some objectives of study in depth are well within the range of an effective joint major, say, in mathematics and biology where senior students employ mathematical models based primarily on lower division mathematics to model a biological phenomenon and then test and modify the model based on laboratory data.

Teacher education poses a special case of particular significance, since mathematics is one of the few disciplines taught throughout all twelve grades of school. It is obviously important for our nation that school teachers be both competent and enthusiastic about mathematics. Special committees recommend standards for preparation of mathematics teachers [14, 18], and these recommendations provide one particular perspective on study in depth.

Prospective secondary school teachers of mathematics generally pursue an undergraduate degree that includes a major in mathematics, often constrained in special ways to ensure breadth appropriate to the responsibilities of high school mathematics teachers. However, the appropriate mathematical preparation of prospective elementary and middle school teachers—who commonly teach several subjects, and sometimes teach the whole curriculum—is subject to much debate these days. Many national studies have recommended that prospective elementary school teachers, like secondary school teachers, major in a liberal art or science rather than in the discipline of education. However, the traditional mathematics major is generally inappropriate for teachers at this level, and today there appears to be virtually no example of a viable alternative. Even more vexing is the question of achieving depth in mathematics appropriate to an elementary school teacher within a major in some other field. Some interesting ideas can be found in the "new liberal arts" initiative sponsored by the Alfred P. Sloan Foundation which has attempted to infuse quantitative methods in traditional liberal arts subjects [38].

Most of the issues, guidelines, and recommendations in this study focus on the traditional mathematics major, which is where most students who study mathematics in depth are to be found. However, study in depth can be done at any level and in many contexts. *Mathematics departments should take seriously the need to provide appropriate mathematical depth for students who wish to concentrate in mathematics without pursuing a traditional major.*

Self-Esteem

One of the greatest impediments to student achievement in mathematics is the widespread reputation of mathematics as a discipline for geniuses. Many facets of school and college practice conspire to portray mathematics in "macho" terms: only those who are bright, aggressive, and inclined towards arrogance are likely to succeed. Those who do not instantly understand—including many thoughtful, reflective, creative students—are made to feel "deeply dumb," like outsiders who don't get the point of an in-joke.

It is hard to overstate the power of intimidation to erode students' self-confidence. Many calculus teachers recognize the problem: bright freshmen "show-offs"—usually white males—whose questions are designed not so much to elicit answers and build understanding as to demonstrate their superior intelligence to their classmates. The ritual is not unlike the bluffing maneuvers that male animals employ to claim dominant status in a herd. Many who are concerned about equality of opportunity believe that the widespread display of "genius-ism" as a measure of worth in mathematics is in part a mask for sexism—an unconscious emphasis on behavior intended to preserve the *status quo* regarding access to leadership in teaching and research.

Fortunately there is a growing recognition in the mathematical community that old traditions must be replaced with new approaches better suited to the demographic realities of our age. We need to recognize that individuals bring very different but equally valuable strengths to the study of mathematics. A multiplicity of approaches that encourage student growth in many different dimensions is far more effective than a single-minded focus leading to a linear ranking in one narrow dimension of "brightness." Not every value in mathematical talent can be measured well by timed tests or intercollegiate competitions; the "Putnam powerhouse" is not the only standard by which undergraduate majors should be judged. (The William Lowell Putnam Examination, a national contest for undergraduates, is the

Nobel competition of collegiate mathematics. It stumps even faculty with questions so hard that the median national score for undergraduates is frequently 0.)

Specific efforts to focus the mathematics curriculum on the interests and abilities of all students can bring dramatic results, as this campus report shows:

> At the time of the first registration for first-year students, fewer than 20 individuals in the entire freshman class indicate that mathematics is a possible major. A year later, the number is in the 50's, and by the junior year the number is over 100. One reason for this impressive increase in student interest in a mathematics major is the departmental position that mathematics is for everyone, not just the gifted. We attempt to demonstrate the power and applicability of mathematics by emphasizing breadth of study during the second and third years.

Self-confidence increases when students succeed, and decreases when they fail. "What students need to build self-confidence are genuine *small successes* of their own" [67]. Initial successes come from routine homework, but these are insufficient to the task. More effective are instructional strategies that engage the student in active learning: open-ended problems, team work that builds diverse problem-solving skills; undergraduate research experiences; independent study. *Building students' well-founded self-confidence should be a major priority for all undergraduate mathematics instruction.*

Access

Data from many sources [3, 48, 65] show that women and members of certain minority groups often discontinue their study of mathematics prematurely, before they are prepared appropriately for jobs or further school. Afro-American and Hispanic students drop out of mathematics at very high rates throughout high school and college; only a tiny fraction complete an undergraduate mathematics major [46]. In college, women major in mathematics almost as often as men do, but they persist in graduate studies at much lower rates. (Interestingly, mathematics comes closer to achieving an even balance of men and women among its undergraduate majors than virtually any other discipline; this record of equality disappears, however, in graduate school.)

Evidence from various intervention programs shows that the high drop-out rate among minority students can be reduced [4, 5, 30, 63]. Appropriate expectations that provide challenges without the stigma of "remediation" together with assignments and study environments that reinforce group learning have proved successful on many campuses. Mentoring programs of various types open doors of opportunity to women and minorities who have traditionally been under-represented in mathematically based fields. What becomes clear from these programs is that the tradition of isolated, competitive individual effort that dominates much mathematics instruction does not provide a supportive learning environment for all students.

Assignments that stress teamwork on problems chosen to relate to student interests can help many students succeed in mathematics. The experiences of students who work in teams to solve large computer science projects and of those who participate in science research groups show clearly the benefit of incentives for careful work that is created by the team atmosphere. Mathematicians must learn that the teaching strategies they recall as being successful in their own education—and in the education of a mostly white male professional class—do not necessarily work as well for those who grow up in vastly different cultures within the American mosaic.

Programs that work for minority students are built on the self-evident premise that students do not all learn mathematics in the same way. Classroom methods must fit both the goals of the major (e.g., to help students to learn to communicate mathematically) and the learning styles of individual students (e.g., need for peer support and positive feedback). These same principles apply to all students, not just to students of color. *To provide effective opportunities for all students to learn mathematics, colleges must offer a broader spectrum of instructional practice that is better attuned to the variety of students seeking higher education.*

Communication

College graduates with majors in mathematically-based disciplines are often perceived by society as being verbally inept: the stereotype of the computer hacker who cannot communicate except with a computer has permeated the business world, and tainted mathematics graduates with the same reputation. Recognizing the legitimate basis for this concern in the incomprehensible writing of their own upper-division students, many mathematics departments are beginning to emphasize writing in mathematics courses at all levels.

The forms of writing employed in mathematics courses include the standard genres used in other disciplines (expository essays, personal journals, laboratory reports, library papers, research reports) as well as some that are particularly relevant to mathematics (proofs, computer programs, solutions to problems). Many students and professors are uneasy about what writing means in a mathematics class, about how to grade it, and how to improve it. Few mathematicians know how to teach students to improve their writing or speaking, although there is increasing professional interest in this issue [31, 37, 61].

One department focuses on communication throughout the major, and stresses writing and speaking mathematics in a required senior colloquium:

> The conclusion of the major features the colloquium course "Mathematical Dialogues." The emphasis here, as in earlier courses, is on communication, as well as on the connections among the different branches of mathematics. Mathematical Dialogues consists of lectures from invited scholars, discussions, and independent work. Students are expected to read papers and write reviews, to listen to talks and to deliver them.

In industry, one of the most important tasks for a mathematician is to communicate to non-mathematicians in writing and orally the mathematical formulation and solution of problems. Each student's growth in mathematical maturity depends in essential ways on continual growth in the ability to communicate in the language of mathematics: to read and write, to listen and speak. Students must learn the idioms of the discipline, and the relation of mathematical symbols to English words. They need to learn how to interpret mathematical ideas arising in many different sources, and how to suit their own expression of mathematics to different audiences. *Mathematics majors should be offered extensive opportunities to read, write, listen, and speak mathematical ideas at each stage of their undergraduate study.* Indeed, writing and speaking is the preferred test of comprehension for most of the broad goals of study in depth.

Transitions

As students grow in mathematical maturity from early childhood experiences to adult employment, they face a series of difficult transitions where the nature of mathematics seems to change abruptly. These "fault lines" that cross the terrain of mathematics education appear at predictable stages:

- Between arithmetic and algebra, when letter symbols, variables, and relationships become important.
- Between algebra and geometry, when logical proof replaces calculation as the methodology of mathematics.
- Between high school and college, when the expectation for learning on one's own increases significantly.
- Between elementary and upper-division college mathematics, when the focus shifts from techniques to theory, from solving problems to writing proofs.
- Between college and graduate school, when the level of abstraction accelerates at a phenomenal rate.
- Between graduate school and college teaching, when the realities of how others learn must take precedence.
- Between graduate school and research, when the new Ph.D. must not just solve a serious problem, but learn to find good problems as well.

Students experience real trauma in crossing these transitions; many drop out of the mathematics pipeline as a consequence, often to the detriment of their future study in many disciplines. Mathematics education at all levels, from grade school through graduate school, should take as a goal to smooth out the roughness caused by these difficult transitions. College mathematics departments should, in particular, seek to streamline the transition of students to college, to upper-division mathematics, and to graduate school.

In college, students often experience a different type of transitional problem that applies in virtually all courses: to understand the relation between theory and applications. This is probably the most common complaint that students and faculty in collateral disciplines raise about undergraduate mathematics courses: they are often perceived as being too theoretical and insufficiently applied. Although in some cases this perception may be well justified, in many other instances the problem rests more with insufficient effort to demonstrate the value of theory to application than with an actual excess of theory. The problem is not that the transition from application to theory is inappropriate, but that it is often taken without sufficient effort to build appropriate motivation or connections. *Smooth curricular transitions improve student learning and help maintain momentum for the study of mathematics.*

Research

The role of so-called "capstone experiences" such as undergraduate research, theses, or senior projects is one of the more controversial ingredients in discussions of the mathematics major. Typically, such requirements are common in the humanities and the sciences, especially in more selective institutions. In the humanities they are viewed as opportunities for integration; in the sciences, as opportunities for research. In both science and humanities, capstone requirements offer apprenticeships in the investigative methods of the field.

In mathematics, however, there has been little consensus about objectives, feasibility, or benefits of this type of requirement. Very few institutions heeded the 1981 CUPM call for a required course in mathematical modelling for all majors. Many mathematicians believe in coverage as more crucial to understanding: standard theorems, paradigms of proof, and significant counter-examples in all major areas must be covered before a student is ready to advance to the next stage of mathematical maturity. In this view, learning what is already known is a prerequisite to discovering the unknown. Moreover, special capstone courses appear superfluous since each course provides its own capstone—the fundamental theorem of calculus, the central limit theorem in statistics, the fundamental theorem of algebra—which ties together a long chain of prior study. When forced to choose between a capstone experience or another advanced course, advocates of coverage will unhesitatingly choose the latter.

Because of mathematics' austere definition of "research"—a definition which, incidentally, rules out the professional work of more than half the nation's mathematics faculty—many mathematicians believe that except in very rare cases, undergraduates cannot do research in mathematics. Moreover, in most areas of mathematics, students cannot even assist in faculty research, as they do quite commonly in the laboratory sciences. The exceptions in mathematics are principally where computer investigation—the mathematician's laboratory—can aid the research effort. As a consequence, many mathematicians believe that further coursework would better serve the goals of integration (because the higher one progresses in mathematics, the more internal links one can see) and at the same time help advance the student towards better preparation for further study or application of mathematics.

Others feel that any encounter with a substantial problem that a student does not know how to solve can provide a legitimate and rewarding research experience. Indeed, many colleges have used summer experiences with undergraduate research as an effective strategy to recruit students to careers in the mathematical sciences [26, 56], and the National Science Foundation is actively supporting such programs. There are now many diverse programs offering research experiences for undergraduate mathematics majors.

In applied areas—especially in statistics, computing, and operations research—it is easier to develop projects that are sufficiently rich and varied so that students can make progress along various lines of investigation. Computers now are making inroads in theoretical areas of mathematics, permitting exploration of conjectures that heretofore were beyond the range of any undergraduate. Students preparing to teach mathematics in high school also have open an enormous range of appropriate projects to translate interesting newer mathematics into curriculum appropriate to the schools. In some cases students may want to undertake research into how people learn mathematics, to explore for themselves the effectiveness of various instructional strategies and the impact of computers on development of mathematical understanding.

Internships in industry, co-op programs that mix study with work, and summer research opportunities in industrial or government laboratories provide rich environments for breaking down the artificial barriers of courses and classrooms. They enable students to integrate mathematics learned in several different courses; to experience the role of mathematical models; to extend their mathematical repertoire beyond just what has been taught; and to establish mathematical concepts in a context of varied use, applications, and connections.

Experiences of departments with long-standing traditions of undergraduate research or senior projects confirm both the value of such work and the effort required for success:

> While these projects require a great deal of time and effort on the part of students and faculty, we generally feel that it is well worth it. Most of the students report that they had worried about the senior project for their first three years, but had ultimately found it to be a very worthwhile

and stimulating part of their college experience. All recommended that this important aspect of the undergraduate experience be retained.

One department in an institution whose academic calendar permits extended blocks for full-time study in one subject requires all majors to complete a major project in the senior year:

> The final project in the major field should demonstrate application of the skills, methods, and knowledge of the discipline to the solution of a problem that would be representative of the type to be encountered in one's career. Project activities encompass research, development and application, involve analysis or synthesis, are experimental or theoretical, emphasize a particular subarea of the major or combine aspects of several subareas.

Another department uses summers to provide opportunities for research experiences: student participants range from freshmen to seniors, and engage in a wide variety of mathematical investigations:

> We are convinced that everyone working in mathematics can find problems appropriate for undergraduates. Many problems can be attacked without any knowledge of the complex machinery which generated them. Mathematicians know how exciting mathematical research can be. The best way to generate interest in mathematics is to provide undergraduates with the chance to experience that excitement.

Since hard work by itself is insufficient to ensure reasonable progress on a mathematical problem, there is ever-present danger that undergraduates confronted with difficult theoretical problems will flounder and become discouraged. Strong faculty intervention can prevent disaster, but excessive supervision undermines the independence that is supposed to result from the project. Effective undergraduate research experiences require careful planning and steady, unobtrusive leadership. One must carefully choose problems to be suggested to undergraduates for the research experiences: they must be tailored to the individual undergraduate.

Effective programs provide stepping stones to help students progress from routine homework to independent investigation. For example, one institution plans a progression leading to the senior project:

> Mathematics majors enroll in a Junior Seminar where they are asked to read critically two senior projects from earlier years to describe the strengths and weaknesses of these papers, and to suggest how they would improve on these papers had they written them. This Seminar also helps acquaint these students with appropriate standards of exposition in mathematics.

The range of opportunities for independent investigation is so broad and the evidence of benefit so persuasive as to make unmistakably clear that research-like experiences should be part of every mathematics student's program. *Undergraduate research and senior projects should be encouraged wherever there is sufficient faculty to provide appropriate supervision.* Effective programs must be tailored to the needs and interests of individual students; no single mode of independent investigation can lay claim to absolute priority over others. Flexibility of implementation is crucial to ensure that all experience the exhilaration of discovery which accompanies involvement with mathematical research.

Context

Mathematics courses—especially those taken by majors—have traditionally been taught as purely utilitarian courses in techniques, theory, and applications of mathematics. Most courses pay no more than superficial attention to the historical, cultural, or contemporary context in which mathematics is practiced. Today, however, as mathematical models are used increasingly for policy and operational purposes of immense consequence, it is vitally important that students of mathematics learn to think through these issues even as they learn the details of mathematics itself.

Examples abound of mathematical activity that leads directly to decisions of great human import. Software written for the Strategic Defense Initiative depends on mathematical theories of orbital dynamics for its performance, and on the ability of logicians and computer scientists to verify that complex untestable programs will perform correctly under any possible situation. Debates about the relation of carbon dioxide build-up to global warming and consequent implications for governmental and industrial policies center in large part on different interpretations of statistical and mathematical projections. Computer-controlled trading of stocks, epidemiological studies of AIDS, and implications of various voting rules offer other examples where mathematics really matters in important decisions affecting daily life.

Students of mathematics should be encouraged to see mathematics as a human subject whose theories often begin in ambiguity and controversy. It takes decades, sometimes centuries, for scholars to sculpt and polish the precise theories that are expounded in today's textbooks. Historical analogs provide useful yardsticks to students (and faculty) who seek to understand the limits of what mathematics can contribute to public policy. As society comes to rely increasingly on mathematical analyses—often well-disguised—of social, economic, or political issues, mathematics majors must confront the social and ethical implications of such activity. All such issues can be enlightened by appropriate historical case studies, and motivated by compelling debates of our age. *All mathematics students should engage in serious study of the historical context and contemporary impact of mathematics.*

One possible strategy to achieve both this objective and several others as well is to adapt a modelling project or course to problems of significant societal impact. In such a setting students could undertake original investigation, gain experience in reading, writing, listening, and speaking about mathematically rich material, explore historical antecedents and contemporary debates, and gain experience in team work to address complex, open-ended problems. For many students a capstone project on a public policy issue would be a fitting way to relate their mathematics major to liberal education.

Social Support

The abstract, austere nature of mathematics provides relatively few intrinsic rewards for the typical undergraduate who is trying to pursue a field of study and at the same time learning to establish and maintain personal friendships. In this context the social support provided by departmental activities can be decisive in tipping the balance either for or against a mathematics major. Peer group support helps build mathematical self-confidence and enhances the intrinsic rewards that come from mathematical achievement.

Virtually all successful mathematics departments instigate and support a variety of extracurricular activities. Examples include mathematics clubs, student chapters of the Mathematical Association of America, or chapters of the mathematics honorary society Pi Mu Epsilon. Another common feature of successful departments is informal faculty-led sessions to help students solve problems posed in collegiate periodicals or to prepare for national contests such as the Putnam Examination or the Mathematical Modelling Contest. Banquets, picnics, and barbecues lend a light touch that help students become acquainted with each other and with the faculty of the entire department.

Other activities can enrich students' experiences with their courses by providing links to the world beyond the campus. Undergraduate colloquia with visiting mathematicians from industry or universities is one common mechanism. Alumni involvement through career nights or other activities can help students imagine what they too could do with their major. Current students will be inspired when departments make visible the variety of accomplishments of their graduates—not only those who have become mathematicians but also the majority who have used their undergraduate mathematics for other ends. *Mathematics departments should exert active leadership in promoting extracurricular activities that enhance peer group support among mathematics majors.*

Mechanisms for Renewal

Constant vigilance is needed to maintain quality in any academic department. This is especially true in mathematics, where the subject is continually evolving, where external departments impose their own often-conflicting demands, where so much teaching effort is devoted to remedial, elementary, and lower-division work, and where the very ability of the discipline to attract sufficient numbers of students to careers in the mathematical sciences is now in serious doubt. We focus here on five mechanisms of renewal:

- DIALOGUE: To talk with students and colleagues.
- ASSESSMENT: To measure what is happening.
- FACULTY DEVELOPMENT: To improve intellectual vitality.
- DEPARTMENTAL REVIEW: To listen to colleagues and clients.
- GRADUATE EDUCATION: To provide leadership for improvement.

The key ingredient is listening—to one's students, to one's discipline, to one's colleagues, to one's friends, and to one's critics. Departments that listen—and learn—will thrive.

Dialogue

Departments often know very little about their students' views of the undergraduate mathematics major. That different students pursue mathematics for very different reasons is clear. Most departments must accommodate students with quite different purposes, although certain departments tend to focus their programs on one or another objective (for example, preparation for jobs, preparation for teaching, preparation for graduate school). Many departments, especially small departments, find it impossible to sustain several different programs of equal high quality.

Mathematicians also frequently know almost nothing about the expectations held by their colleagues in cognate disciplines for the mathematical preparation of students with other majors. It is not uncommon for the three interested parties—mathematics professors, science faculty advisors, and students—never to discuss goals or objectives, but only credit hour requirements. It should come as no surprise that in the absence of good communication, misunderstandings flourish.

Undergraduate mathematics shares many borders with other subjects and institutions: vertically with high schools below and graduate schools above; horizontally with science, business, and engineering. Each border is a potential impediment to the smooth flow of ideas and students. Mathematics departments must work hard to maintain effective articulation across these many boundaries:

- With high schools whose curriculum is also changing and whose students will arrive at college with new expectations.
- With departments in the physical sciences and engineering whose students use advanced mathematics.
- With graduate schools in the mathematical sciences, which attract and retain far too few U.S. students.
- With employers of bachelors degree graduates who expect employees who can function effectively in a work environment.

Regular discussion is essential to maintain effective policies that will satisfy these many boundary conditions.

To the extent that resources permit, departments should seek to determine and then accommodate different student career interests. This means that even small departments should provide mechanisms (e.g., independent study, special seminars) to allow students of diverse interests to receive a major suitable to their career objectives. Mathematics is too diverse and student purposes too different for any single set of eight to ten courses to meet all needs equally well.

Students too must recognize that the practice of mathematics is quite different from the textbook image they usually bring with them from high school. Often students expect of college mathematics merely advanced topics in the spirit of school mathematics: a succession of techniques, exercises, and test problems, each explained by the instructor with sufficient clarity that what remains for the student is only the requirement of practice and memorization. Such expectations do little to foster creativity, independence, criticism, and perspective—the more important goals of liberal education.

The different perspectives of mathematics student and mathematics professor often approach caricature. Eager students expect of college classes directed instruction in tools of the trade with which they can, upon graduation, get jobs that pay more than their professors earn. Professors, in contrast, expect students who are eager to take on challenging problems and who will learn on their own whatever they need to make progress. Students, in this exaggerated portrait, feel responsible only for what they have been taught, whereas faculty judge as truly significant only those things students can do which they have not been taught.

It is important for mathematics departments to help faculty and students recognize their own perspectives on mathematics and understand the perspectives of others. Doing this is not the same as covering a syllabus of mathematical topics; it involves instead various strategies to enable faculty and students to discuss mathematics in informal ways. Such discussions are an important part of the process by which students grow from the limited school perspective to the self-directed stance of a professional.

Announcing or publishing department goals is not sufficient to achieve this important objective. What is required is a process that engages all students in significant and repeated discussion of individual goals throughout their undergraduate study of mathematics. In particular, *careful and individualized advising is crucial to students' success.* Effective advising builds an atmosphere of mutual respect among faculty and students. Courses, career objectives, motivations, fears, celebrations are all part of advising, and of special importance in the long, slow process of building students' self-confidence.

Assessment

Many would argue that goals for study in depth can be effective only if supported by a plan for assessment that persuasively relates the work on which students are graded to the objectives of their education. Assessment in courses and of the major as a whole should be aligned with appropriate objectives, not just with the technical details of solving equations or doing proofs. Many specific objectives can flow from the broad goals of study in depth, including solving open-ended problems; communicating mathematics effectively; close reading of technically-based material; productive techniques for contributing to group efforts; recognizing and expressing mathematical ideas embedded in other contexts. Open-ended goals require open-ended assessment mechanisms; although difficult to use and interpret, such devices yield valuable insight into how students think.

Relatively few mathematics departments now require a formal summative evaluation of each student's major. The few that do often use the Graduate Record Examination (or an undergraduate counterpart) as an objective test, together with a local requirement for a paper, project, or presentation on some special topic. Many institutions, frequently pressured by mandates from on high, are developing comprehensive plans for assessing student outcomes; a few are exploring innovative means of assessment based on portfolios, outside examiners, or undergraduate research projects. Here's one example that blends a capstone course with a senior evaluation:

> The Senior Evaluation has two major components to be completed during the fall and spring semesters of the senior year. During the fall semester the students are required to read twelve carefully selected articles and to write summaries of ten of them. (Faculty-written summaries of two articles are provided as examples.) This work comprises half the grade on the senior evaluation. During the fall semester each student chooses one article as a topic for presentation at a seminar. During the spring semester the department arranges a seminar whose initial talks are presented by members of the department as samples for the students. At subsequent meetings, the students present their talks. Participation in the seminar comprises the other half of the grade for the Senior Evaluation.

Because of the considerable variety of goals of an undergraduate mathematics major, it is widely acknowledged that ordinary paper-and-pencil tests cannot by themselves constitute a valid assessment of the major. Although some important skills and knowledge can be measured by such tests, other objectives (e.g., oral and written communication; contributions to team work) require other methods. Some departments are beginning to explore portfolio systems in which a student submits samples of a variety of work to represent just what he or she is capable of. A portfolio system allows students the chance to put forth their best work, rather than judging them primarily on areas of weakness.

The recommendations [41] from the National Council of Teachers of Mathematics for evaluation and assessment of school mathematics convey much wisdom that is applicable to college mathematics. Assessment must be aligned with goals of instruction. If one wants to promote higher order thinking and habits of mind suitable for effective problem solving, then these are the things that should be tested. Moreover, assessment should be an integral part of the process of instruction: it should arise in large measure out of learning environments in which the instructor can observe how students think as well as whether they can find right answers. *Assessment of undergraduate majors should be aligned with broad goals of the major: tests should stress what is most important, not just what is easiest to test.*

Faculty Development

The relation of research and scholarship to faculty vitality is one of the most difficult issues facing many departments of mathematics, especially in smaller institutions. Professional activity is crucial to inspired teaching and essential to avoid faculty burn-out. Mathematical research in its traditional sense plays only a small role in the mechanisms required to maintain intellectual vitality of a mathematics department: only about one in five full-time faculty in departments of mathematics publish regularly in research journals, and fewer than half of those have any financial support for their research. Clearly the community needs to encourage and support a broader standard as a basis for maintaining faculty leadership both in curriculum and in scholarship.

The first step is to expand the definition of professional activity from "research" to "scholarship," more in a manner akin to that currently recognized in some other academic disciplines. Applied consulting work, software development, problem solving, software and book reviews, expository writing, and curriculum development are examples of activities that serve many of the same purposes as research: they advance the field in particular directions, they engage faculty in active original work, they serve as models for students of how mathematics is actually practiced, and they provide opportunities for student projects.

Teaching in new areas is also a form of scholarship in mathematics. Unlike many other disciplines where faculty rarely teach outside their own areas of specialty, mathematicians are generally expected to teach a wide variety of courses. Learning and then teaching a course far outside one's zone of comfort is an effective way to build internal connections which then spill over in all courses one teaches. A teacher who is still an active learner sets a fine example for students concerning the true meaning of scholarship.

The second step is to insist on greater communication about professional activity in mathematics so that it becomes public. Only the bright light of public scrutiny by colleagues in various institutions—not only on one's own campus—can affirm the quality and value of professional work. "Public" need not mean merely publication; lectures, workshops, demonstrations, reports of various sorts can serve the same objective. What matters is that the result become part of the profession, and be evaluated by the profession. *To ensure continued vitality of undergraduate mathematics programs, all mathematics faculty should engage in public professional activity, broadly defined.*

Departmental Review

More than any other academic discipline, mathematics is constrained to serve many masters: the many sciences that depend on mathematical methods; the demand of quantitative literacy that undergirds general education; the need to educate teachers for our nation's schools; the need of business and industry for mathematically literate employees; the expectation of mathematical proficiency by faculty and students in natural science, business, engineering, and social science; the professional standards of employers for entry-level technical personnel; and the requirements of the mathematical sciences themselves for well-prepared graduate students. It is an enormous challenge for a department of mathematics, one that very few are able to fulfill with distinction in every dimension.

Because of these diverse demands, it is especially important that departments of mathematics undergo regular review, with both external and internal mechanisms to provide evaluation and advice. External requirements mandate periodic review of all departments in many colleges and universities, especially in public institutions. But in other institutions, department goals are defined implicitly without self-reflection or benefit of external perspectives. At worst, the goals of such departments are defined by coverage of standard textbooks. Often it takes a crisis—such as when the engineering or business school complains about certain courses—for departments to step back and examine their objectives. Reviews should take place regularly, not just when some crisis threatens the *status quo*.

Client disciplines expect from mathematics departments an amazing repertoire of support services for students who will major in other fields [44]. Some demand a magic bullet—a perfect infusion of just those mathematical methods (and no more) needed in the other field; others expect a rigorous filter that will pass on only those students who are sufficiently bright to function ably in upper-division work in other fields. Occasionally, but all too rarely, an external discipline will require mathematics primarily to enable students to benefit from the intrinsic values of mathematics: logic, rigor, analysis, symbol-sense, etc. Since the expectations of other departments are often not clearly conveyed by the list of mathematics courses that they recommend or require, regular reviews provide a good mechanism—but not the only one—to ensure that different departments at least understand their differing perspectives and objectives.

Virtually all departments receive informal feedback from graduates, employers, and graduate schools. Speaker programs that bring students and faculty into contact with users of mathematics serve both to inform students about the broad world of mathematics beyond their classroom walls and to provide informal feedback to help regulate the curriculum and keep it properly tuned to the needs of graduates. All such informal means of feedback are valuable and must be encouraged. However, they are no substitute for formal, regular, external review. *Both external reviews and informal feedback are needed to assure quality in departments of mathematics.*

There are many advantages to a regular program of external reviews that should form the basis of all reviews:

- A broad-based review provides a strategic opportunity to document the accomplishments of a department. Well-structured reviews can effectively counter external (political) demands for narrow or inappropriate instruments of assessment such as a multiple-choice examination of all graduates.

- Reviews provide a structured and neutral forum for mathematicians to discuss with those who use mathematics both the mathematical needs of client disciplines and the common issues that both mathematics and the client discipline face in accommodating changes that are underway in the mathematics curriculum.

- By involving members of the faculty outside the department in the review—especially those from fields that are served by mathematics and those involved in faculty curriculum committees—a department of mathematics can help educate colleagues across the campus about the special opportunities and challenges of teaching mathematics. Ignorance can usually be turned to understanding through discussions prompted by the occasion of a regular review.

- By including non-academic reviewers such as industrial executives, scientists, professionals, and community leaders, the department can gain valuable insight into the qualities that will be expected of graduates who enter the work force.

- Regular reviews encourage faculty members to think about the department's program as a whole, rather than only about the courses they teach. Such discussions make it more likely that the curriculum will remain responsive to student needs, and to the changing demands of the mathematical sciences. Reviews provide an ideal mechanism for the department to assert control over its own program.

The Mathematical Association of America can provide advice to departments both about the structure of effective reviews and about appropriate consultants or reviewers.

Graduate Education

Even though relatively few mathematics majors go on to receive a graduate degree in the mathematical sciences, the health of college mathematics is inextricably linked with the status of graduate education. As the sole agent for advanced degrees, graduate schools bear alone the responsibility for preparing college mathematics teachers; as the primary locus of mathematical research, graduate schools shape the nature of the discipline, and hence of the curriculum. Much of the responsibility for renewing undergraduate mathematics rests with the graduate schools, since it is they who provide the primary professional education of those who are responsible for undergraduate mathematics: college faculty.

Indicators from many sources [12, 35, 44] suggest that the match between undergraduate and graduate education in mathematics is not now serving U.S. interests especially well:

- Too few U.S. mathematics majors choose to enter graduate school in a mathematical science.
- U.S. mathematics students do less well in graduate school—and drop out more often—than foreign nationals.
- Many students finish graduate school ill-equipped for the breadth of teaching duties typically expected of undergraduate mathematics teachers.
- Relatively few who finish doctoral degrees in mathematics actually go on to effective research careers in mathematics.

In the 1970s, as the number of U.S. students applying to graduate school in mathematics began to decline, the graduate schools responded by increasing the number of international students, most of whom had completed a more intense and specialized education in mathematics than is typical of American undergraduates. Hence the level of mathematics expected of beginning graduate students gradually shifted upward to an international standard that is well above current U.S. undergraduate curricula. Consequently, the failure or drop-out rate of U.S. students increased, creating pressure for more international students and even higher entrance expectations.

It is time to break this negative feedback loop by encouraging better articulation of programs and standards between U.S. undergraduate colleges and U.S. graduate schools. Such cooperation is needed both to enhance the success of U.S. students and to enable the graduate schools to better match their programs with the needs of the colleges and universities who employ a majority of those who receive advanced degrees. *Renewal of undergraduate mathematics will require commitment, leadership, and support of graduate schools.*

One good mechanism for such cooperation would be an exchange of visitors between undergraduate and graduate institutions so that each can learn about the needs of the other. Especially as change occurs in the content and nature of the undergraduate major, it is very important that graduate schools maintain programs of study and research that are appropriately linked to the undergraduate program in mathematics.

Summary

Without becoming entangled in specific curriculum and course recommendations—which are the proper province of other committees of mathematics professional organizations—we can nevertheless enumerate several broad principles implied by our study of the undergraduate mathematics major:

GOALS AND OBJECTIVES

- The primary goal of a mathematical sciences major should be to develop a student's capacity to undertake intellectually demanding mathematical reasoning.
- The undergraduate mathematics curriculum should be designed for all students with an interest in mathematics.
- Applications should motivate theory so that theory is seen by students as useful and enlightening.
- Mathematics majors should be offered extensive opportunities to read, write, listen, and speak mathematical ideas at each stage of their undergraduate study.

BREADTH AND DEPTH

- All students who major in mathematics should study some sequence of upper division courses that shows the power of study in depth.
- Every student who majors in mathematics should study a broad variety of advanced courses.
- Mathematics departments should take seriously the need to provide appropriate mathematical depth to students who wish to concentrate in mathematics without pursuing a traditional major.
- Mathematics majors should complete a minor in a discipline that makes significant use of mathematics.

LEARNING AND TEACHING

- Instruction should encourage students to explore mathematical ideas on their own.
- Undergraduate students should not only learn the subject of mathematics, but also learn how to learn mathematics.
- Those who teach college mathematics should seek ways to incorporate into their own teaching styles the findings of research on teaching and learning.
- Mathematicians should increase their efforts to understand better how college students learn mathematics.
- Evaluation of teaching must involve robust indicators that reflect the broad purposes of mathematics education.

ACCESS AND ENCOURAGEMENT

- Effective programs teach students, not just mathematics.
- National need requires greater encouragement for students to continue their study of mathematics beyond the bachelor's degree.
- To provide effective opportunities for all students to learn mathematics, colleges should offer a broader spectrum of instructional practice that is better attuned to the variety of students seeking higher education.
- To ensure for all students equal access to higher mathematics education, mathematics departments should work with nearby two-year colleges to maintain close articulation of programs.
- Smooth curricular transitions improve student learning and help maintain momentum for the study of mathematics.

USING COMPUTERS

- The mathematics curriculum should change to reflect in appropriate ways the impact of computers on the practice of mathematics.
- Colleges must recognize in budgets, staffing, and space the fact that undergraduate mathematics is rapidly becoming a laboratory discipline.

DOING MATHEMATICS

- Dealing with open-ended problem situations should be one of the highest priorities of undergraduate mathematics.
- All undergraduate mathematics students should undertake open-ended projects whose scope extends well beyond typical textbook problems.
- Undergraduate research and senior projects should be encouraged wherever there is sufficient faculty to provide appropriate supervision.
- Students majoring in mathematics should undertake some real-world mathematical modelling project.

STUDENTS

- Building students' well-founded self-confidence should be a major priority for all undergraduate mathematics instruction.
- Careful and individualized advising is crucial to students' success.
- All mathematics students should engage in serious study of the historical context and contemporary impact of mathematics.
- Mathematics departments should actively encourage extracurricular programs that enhance peer group support among mathematics majors.

RENEWAL

- It is important for mathematics departments to help faculty and students recognize their own perspectives on mathematics and understand the perspectives of others.
- Assessment of undergraduate majors should be aligned with broad goals of the major; tests should stress what is most important, not just what is easiest to test.
- To ensure continued vitality of undergraduate mathematics programs, all mathematics faculty should engage in public professional activity, broadly defined.

- Regular external reviews and informal feedback are needed to assure quality in departments of mathematics.
- Renewal of undergraduate mathematics will require commitment, leadership, and support of graduate schools.

In most respects both prevailing professional wisdom and current practice for the mathematics major reflect well the major goals of AAC's *Integrity*. Discussion continues on many campuses about whether the major should focus inward towards advanced study in the mathematical sciences or outward towards preparation for diverse careers in science and management. These discussions are more about strategies than long-term goals, however, since either emphasis can advance the broad AAC goals of coherence, connections, and intellectual development.

Liberal education provides a versatile background for a life of ever-changing challenges. Among the many majors from which students can choose, mathematics can help ensure versatility for the future. Habits of mind nurtured in an undergraduate mathematics major are profoundly useful in an enormous variety of professions. The challenge for college mathematicians is to ensure that the major provides—and is seen by students as providing—not just technical facility, but broad empowerment in the language of our age.

References

[1] Albers, Donald J.; Anderson, Richard D.; Loftsgaarden, Don O. *Undergraduate Programs in the Mathematical and Computer Sciences: The 1985-1986 Survey.* MAA Notes No. 7. Washington, DC: Mathematical Association of America, 1987.

[2] Albers, Donald J.; Rodi, Stephen B.; Watkins, Ann E. (Eds.). *New Directions in Two-Year College Mathematics.* New York: Springer-Verlag, 1985.

[3] American Council on Education. *One Third of a Nation.* Report of the Commission on Minority Participation in Education and American Life. Washington, DC: American Council on Education, 1988.

[4] Anderson, Beverly; Carl, Iris; Leinwand, Steven (Eds.). *Making Mathematicians Work for Minorities.* CBMS Issues in Mathematics Education. Providence, R.I.: American Mathematical Society (To Appear).

[5] Asera, Rose. "The Math Workshop: A Description." In Fisher, Naomi; Keynes, Harvey; Wagreich, Philip (Eds.). *Mathematicians and Education Reform.* CBMS Issues in Mathematics Education, Volume 1. Providence, RI: American Mathematical Society, 1990, pp. 47–62.

[6] Association of American Colleges. *Integrity in the College Curriculum.* Washington, DC: Association of American Colleges, 1985.

[7] Board on Mathematical Sciences. *Mathematical Sciences: A Unifying and Dynamic Resource.* National Research Council. Washington, DC: National Academy Press, 1986.

[8] Board on Mathematical Sciences. *Mathematical Sciences: Some Research Trends.* National Research Council. Washington, DC: National Academy Press, 1988.

[9] Board on Mathematical Sciences. *Renewing U.S. Mathematics: A Plan for the 1990s.* National Research Council. Washington, DC: National Academy Press, 1991.

[10] Boyer, Ernest L. *College: The Undergraduate Experience.* New York: Harper & Row, 1987.

[11] Case, Bettye Anne. *Teaching Assistants and Part-time Instructors.* Washington, DC: Mathematical Association of America, 1988.

[12] Case, Bettye Anne. "How Should Mathematicians Prepare for College Teaching?" *Notices of the American Mathematical Society*, 36 (December 1989) 1344–1346.

[13] College Entrance Examination Board. *Academic Preparation in Mathematics: Teaching for Transition From High School To College.* New York: The College Board, 1985.

[14] Committee on the Mathematical Education of Teachers. *A Call for Change: Recommendations for the Mathematical Preparation of Teachers of Mathematics.* Washington, DC: Mathematical Association of America, 1991.

[15] Committee on the Undergraduate Program in Mathematics. *A General Curriculum in Mathematics for Colleges.* Washington, DC: Mathematical Association of America, 1965.

[16] Committee on the Undergraduate Program in Mathematics. *Recommendations for a General Mathematical Sciences Program.* Washington, DC: Mathematical Association of America, 1981.

[17] Committee on the Undergraduate Program in Mathematics. *Reshaping College Mathematics.* MAA Notes No. 13. Washington, DC: Mathematical Association of America, 1989.

[18] Committee on the Undergraduate Program in Mathematics Panel on Teacher Training. *Recommendations on the Mathematical Preparation of Teachers.* MAA Notes No. 2. Washington, DC: Mathematical Association of America, 1983.

[19] Conference Board of the Mathematical Sciences. *New Goals for Mathematical Sciences Education.* Washington, DC: Conference Board of the Mathematical Sciences, 1984.

[20] Connors, Edward A. "A Decline in Mathematics Threatens Science—and the U.S." *The Scientist*, 2 (November 28, 1988) 9, 12.

[21] Connors, Edward A. "1989 Annual AMS-MAA Survey." *Notices of the American Mathematical Society*, 36 (November 1989) 1155–1188.

[22] Cunningham, Steven and Zimmerman, Walter (Eds.). *Visualization in Teaching and Learning Mathematics.* MAA Notes No. 19. Washington, DC: Mathematical Association of America, 1991.

[23] David, Edward E. (Ed.). *Renewing U.S. Mathematics: Critical Resource for the Future.* Committee on Resources for the Mathematical Sciences, National Research Council. Washington, DC: National Academy Press, 1984.

[24] David, Edward E. "Renewing U.S. Mathematics: An Agenda to Begin the Second Century." *Notices of the American Mathematical Society*, 35 (October 1988) 1119–1123.

[25] Davis, Philip J. "Applied Mathematics as Social Contract." *Mathematics Magazine*, 61 (1988) 139–147.

[26] Davis-Van Atta, David, et al. *Educating America's Scientists: The Role of the Research College.* Oberlin, OH: Oberlin College, 1985.

[27] Fisher, Naomi; Keynes, Harvey; Wagreich, Philip (Eds.). *Mathematicians and Education Reform.* CBMS Issues in Mathematics Education, Volume 1. Providence, RI: American Mathematical Society, 1990.

[28] Garfunkel, Solomon and Young, Gail. "A Study of Mathematical Offerings in Diverse Departments." In *Mathematical Sciences: Servant to Other Disciplines.* Committee on Mathematical Sciences in the Year 2000, National Research Council. Washington, DC: National Academy Press (To Appear).

[29] Gilfeather, Frank. "University Support for Mathematical Research." *Notices of the American Mathematical Society*, 34 (November 1987) 1067–1070.

[30] Gillman, Leonard. "Teaching Programs That Work." *Focus*, 10:1 (1990) 7–10.

[31] Gopen, George D. and Smith, David A. "What's an Assignment Like You Doing in a Course Like This?: Writing to Learn Mathematics." In Connolly, Paul and Yilardi, Teresa (Eds.). *Writing to Learn Mathematics and Science.* New York: Teachers College Press, 1989; reprinted in *The College Mathematics Journal,* 21 (1990) 2–19.

[32] Graham, William R. "Challenges to the Mathematics Community." *Notices of the American Mathematical Society,* 34 (February 1987) 245–250.

[33] Howson, Geoffrey; Kahane, J.-P.; Lauginie, P.; de Turckheim, E. (Eds.). *Mathematics as a Service Subject.* International Commission on Mathematical Instruction Study Series. Cambridge: Cambridge University Press, 1988.

[34] Howson, Geoffrey and Kahane, J.-P. (Eds.). *The Influence of Computers and Informatics on Mathematics and Its Teaching.* International Commission on Mathematical Instruction Study Series. Cambridge: Cambridge University Press, 1986.

[35] Jackson, Allyn B. "Graduate Education in Mathematics: Is it Working?" *Notices of the American Mathematical Society,* 37 (1990) 266–268.

[36] Karian, Zaven. *Report of the CUPM Subcommittee on Symbolic Computation.* Washington, DC: Mathematical Association of America (To Appear).

[37] Knuth, Donald E.; Larrabee, Tracy; Roberts, Paul M. *Mathematical Writing.* MAA Notes No. 14. Washington, DC: Mathematical Association of America, 1989.

[38] Koerner, James D. (Ed.). *The New Liberal Arts: An Exchange of Views.* New York: Alfred P. Sloan Foundation, 1981.

[39] Madison, Bernard L. and Hart, Therese A. *A Challenge of Numbers: People in the Mathematical Sciences.* Committee on Mathematical Sciences in the Year 2000, National Research Council. Washington, DC: National Academy Press, 1990.

[40] McKnight, Curtis C., et al. *The Underachieving Curriculum: Assessing U.S. School Mathematics from an International Perspective.* Champaign, IL: Stipes Publishing Company, 1987.

[41] National Council of Teachers of Mathematics. *Curriculum and Evaluation Standards for School Mathematics.* Reston, VA: National Council of Teachers of Mathematics, 1989.

[42] National Institute of Education. *Involvement in Learning: Realizing the Potential of American Higher Education.* Washington, DC: National Institute of Education, 1984.

[43] National Research Council. *Everybody Counts: A Report to the Nation on the Future of Mathematics Education.* Washington, DC: National Academy Press, 1989.

[44] National Research Council. *Moving Beyond Myths: Revitalizing Undergraduate Mathematics.* Committee on Mathematical Sciences in the Year 2000, National Research Council. Washington, DC: National Academy Press, 1991.

[45] National Science Board Task Committee on Undergraduate Science and Engineering Education. *Undergraduate Science, Mathematics and Engineering Education, Vols. I, II.* Washington, DC: National Science Foundation, 1986, 1987.

[46] National Science Foundation. *Women and Minorities in Science and Engineering.* Washington, DC: National Science Foundation, 1988.

[47] Nisbett, Richard E., et al. "Teaching Reasoning." *Science,* 238 (October 30, 1987) 625–631.

[48] Oaxaca, Jaime and Reynolds, Ann W. (Eds.). *Changing America: The New Face of Science and Engineering, Final Report.* Task Force on Women, Minorities, and the Handicapped in Science and Technology, January 1990.

[49] Office of Technology Assessment. *Educating Scientists and Engineers, Grade School to Grad School.* Washington, DC: Office of Technology Assessment, 1988.

[50] Ralston, Anthony and Young, Gail S. *The Future of College Mathematics.* New York: Springer-Verlag, 1983.

[51] Resnick, Lauren B. *Education and Learning to Think.* Committee on Mathematics, Science, and Technology Education; Commission on Behavioral and Social Sciences and Education, National Research Council. Washington, DC: National Academy Press, 1987.

[52] Rheinboldt, Werner C. *Future Directions in Computational Mathematics, Algorithms, and Scientific Software.* Philadelphia, PA: Society for Industrial and Applied Mathematics, 1985.

[53] Romberg, Thomas A. "Policy Implications of the Three R's of Mathematics Education: Revolution, Reform, and Research." Paper presented at annual meeting of American Educational Research Association, 1988.

[54] Schoenfeld, Alan H. *Mathematical Problem Solving.* New York: Academic Press, 1985.

[55] Schoenfeld, Alan H. (Ed.). *A Source Book for College Mathematics Teaching.* Committee on the Teaching of Undergraduate Mathematics. Washington, DC: Mathematical Association of America, 1990.

[56] Senechal, Lester. *Models for Undergraduate Research in Mathematics.* MAA Notes No. 18. Washington, DC: Mathematical Association of America, 1990.

[57] Simon, Barry (Ed.). *Report of the Committee on American Graduate Mathematics Enrollments.* Washington, DC: Conference Board of the Mathematical Sciences, 1987; *Notices of the American Mathematical Society,* 34 (August 1987) 748–750.

[58] Smith, David A.; Porter, Gerald J.; Leinbach, L. Carl; Wenger, Ronald H. (Eds.). *Computers and Mathematics: The Use of Computers in Undergraduate Instruction.* MAA Notes No. 9. Washington, DC: Mathematical Association of America, 1988.

[59] Steen, Lynn Arthur (Ed.). *Calculus for a New Century: A Pump, Not a Filter.* MAA Notes No. 8. Washington, DC: Mathematical Association of America, 1988.

[60] Steen, Lynn Arthur. "Developing Mathematical Maturity." In Ralston, Anthony and Young, Gail S. (Eds.). *The Future of College Mathematics.* New York: Springer-Verlag, 1983, pp. 99–110.

[61] Sterrett, Andrew (Ed.). *Using Writing to Teach Mathematics.* MAA Notes No. 16. Washington, DC: Mathematical Association of America, 1990.

[62] Thurston, William P. "Mathematical Education." *Notices of the American Mathematical Society,* 37 (September 1990) 844–850.

[63] Treisman, Philip Uri. "A Study of the Mathematics Performance of Black Students at the University of California, Berkeley." In Fisher, Naomi; Keynes, Harvey; Wagreich, Philip (Eds.). *Mathematicians and Education Reform.* CBMS Issues in Mathematics Education, Volume 1. Providence, RI: American Mathematical Society, 1990, pp. 33–46.

[64] Tucker, Thomas (Ed.). *Priming the Calculus Pump: Innovations and Resources.* MAA Notes No. 17. Washington, DC: Mathematical Association of America, 1990.

[65] Widnall, Sheila E. "AAAS Presidential Lecture: Voices from the Pipeline." *Science,* 241 (September 30, 1988) 1740–1745.

[66] Wilf, Herbert S. "The Disk with the College Education." *American Mathematical Monthly,* 89 (1982) 4–8.

[67] Wilf, Herbert S. "Self-esteem in Mathematicians." *The College Mathematics Journal,* 21:4 (September 1990) 274–277.

[68] Zorn, Paul. "Computing in Undergraduate Mathematics." *Notices of the American Mathematical Society,* 34 (October 1987) 917–923.

The Undergraduate Major in the Mathematical Sciences

COMMONS
Department of
Mathematic

The Undergraduate Major in the Mathematical Sciences

A Report of The Committee on the Undergraduate Program in Mathematics

Foreword

It is now twenty years since the post-Sputnik boom created a peak of over 25,000 undergraduate mathematics majors per year. It is ten years since CUPM issued its last major report [54] on the undergraduate major, a report that stressed the need to broaden the major's focus from undergraduate *mathematics* to undergraduate *mathematical sciences*. The present document reflects the wisdom of practice that has emerged within mathematical sciences departments over the past decade, and makes important suggestions for new areas of emphasis.

In recent years, the national spotlight has been aimed at mathematics education, both in schools and in colleges. One result has been a coordinated effort by the entire mathematical community to set standards for curriculum, for teaching, and for assessment. Six recent reports—two from NCTM (*Curriculum and Evaluation Standards for School Mathematics* and *Professional Standards for Teaching Mathematics*), two from MAA (*A Source Book for College Mathematics Teaching* and *A Call for Change: Recommendations for the Mathematical Preparation of Teachers of Mathematics*), one from AAC and MAA (*Challenges for College Mathematics: An Agenda for the Next Decade*), and one from NRC (*Moving Beyond Myths: Revitalizing Undergraduate Mathematics*)—have set the stage for significant improvements in undergraduate mathematics.

This new CUPM report on curriculum emerges into a landscape already over-filled with advice, recommendations, and calls for reform. It is, in some respects, the linchpin in the entire system, since no reform can really succeed unless the roots of that effort are deeply planted in effective practices of undergraduate teaching. It is in mainstream courses of the mathematics major that prospective secondary school teachers learn what mathematics is and how it is to be taught; that students decide to advance or abandon their mathematical education; and that prospective science majors either succeed or fail at achieving literacy in the language of science.

This report is not a call for revolution, but an affirmation of the many good changes that have occurred during the past decade and an exhortation to focus especially on certain issues that are crucial to success in undergraduate mathematics. Departments should read it not in isolation, but in the context of the related reports cited above. Mathematics education is a seamless system from grade school through graduate school: improvement in any part requires coordinated and consistent improvement in every part. CUPM hopes that this timely and well-focused report will assist departments in the urgent and important task of making mathematics into a pump for our nation's scientific pipeline.

Lynn Arthur Steen
CUPM Chair
St. Olaf College

Preface

The Mathematical Association of America, through its Committee on the Undergraduate Program in Mathematics, has long given high priority to recommendations about the undergraduate major. This new report, while anchored firmly in the reality of current practice, owes much to previous recommendations on the undergraduate major in the mathematical sciences.

How does this document move forward? The first five of nine tenets of philosophy of the current report are similar to those of the 1981 CUPM report [54], with sharpenings to encourage independent mathematical learning and attention to written and oral communication of mathematics. Four new tenets of philosophy deal with choices of tracks, the resulting increased advising responsibilities, effects and applications of technology, and "pipeline" issues. A unified structure, presented as a tool for fashioning undergraduate departmental course requirements, allows broad course choices satisfying these nine tenets. These recommendations link immediate utility with flexibility to support the changes of a life career.

Acknowledgements. The CUPM Subcommittee on the Major in the Mathematical Sciences was established in January 1987 to "focus on the third and fourth years of the mathematics curriculum and aim to update" *Recommendations for a General Mathematical Sciences Program* [54]. A public forum in 1989, discussions at some Section meetings, and responses to announcements in professional publications generated useful comments from the mathematical community. Communication with related committees was active, and helpful data was provided to the subcommittee. In particular, we thank Richard D. Anderson, Jean Calloway, Edward Connors, John Fulton, James Leitzel, Bernard Madison, Richard Neidinger, Barry Simon, Lynn Steen, and James Voytuk.

During 1990, three special writing sessions were held to produce a series of drafts which were considered by both current and some former members of the Subcommittee, as well as by several outside readers. Feasibility was at all times a priority during the investigation, deliberation, and writing processes. In January 1991 CUPM voted unanimously to endorse the report.

The Chair appreciated the generosity of time and spirit of the committee members who often completed research and writing on short notice; the committee appreciated helpful responses by outside readers. Some travel support was provided by MAA and NSF; Florida State University provided communications and word processing support. Melissa E. Smith formatted the successive drafts in TEX and helped us meet deadlines.

Bettye Anne Case
Subcommittee Chair
Florida State University
December 1991

Committee on the Undergraduate Program in Mathematics (CUPM):

JERRY L. BONA, *Pennsylvania State University.*

BETTYE ANNE CASE, *Florida State University.*

JAMES W. DANIEL, *University of Texas at Austin.*

JAMES DONALDSON, *Howard University.*

WADE ELLIS, JR., *West Valley College.*

JEROME A. GOLDSTEIN, *Tulane University and MSRI.*

DEBORAH T. HAIMO, *University of Missouri, St. Louis.*

BARBARA A. JUR, *Macomb Community College.*

ZAVEN KARIAN, *Denison University.*

BERNARD MADISON, *University of Arkansas.*

WILLIAM A. MARION, JR., *Valparaiso University.*

SEYMOUR V. PARTER, *University of Wisconsin, Madison.*

SHARON CUTLER ROSS, *DeKalb College.*

LESTER J. SENECHAL, *Mount Holyoke College.*

LINDA R. SONS, *Northern Illinois University.*

IVAR STAKGOLD, *University of Delaware.*

LYNN ARTHUR STEEN, CHAIR, *St. Olaf College.*

URI TREISMAN, *University of California, Berkeley.*

ALAN C. TUCKER, *SUNY at Stony Brook.*

THOMAS TUCKER, *Colgate University.*

CAROL WOOD, *Wesleyan University.*

Subcommittee on the Major in the Mathematical Sciences (SUM):

RICHARD A. ALO, *University of Houston, Downtown.*

JERRY L. BONA, *Pennsylvania State University.*

JEAN M. CALLOWAY, *Kalamazoo College.*

BETTYE ANNE CASE, CHAIR, *Florida State University.*

J. KEVIN COLLIGAN, *National Security Agency.*

JAMES DONALDSON, *Howard University.*

JEROME A. GOLDSTEIN, *Tulane University and MSRI.*

KENNETH I. GROSS,* *University of Vermont.*

JAMES M. HYMAN, *Los Alamos National Labs.*

ELEANOR G. DAWLEY JONES, *Norfolk State University.*

PETER D. LAX,* *Courant Institute, New York University.*

DAVID J. LUTZER, *College of William and Mary.*

WILLIAM F. LUCAS, *Claremont Graduate School.*

WILLIAM A. MARION, *Valparaiso University.*

RONALD J. MIECH,* *University of California at Los Angeles.*

SAMUEL M. RANKIN III, *Worcester Polytechnic Institute.*

MICHAEL C. REED,* *Duke University.*

ALAN C. TUCKER, *SUNY at Stony Brook.*

CAROL WOOD,* *Wesleyan University.*

GAIL S. YOUNG,* *Teachers College, Columbia University.*

* *Term on Subcommittee ended before completion of the report*

The Undergraduate Major in the Mathematical Sciences

Introduction

As the 21st century approaches, better mathematics preparation and involvement of more students in higher levels of achievement are recognized as central components in efforts to remain competitive in the world economy. The collegiate mathematics community, aware of the foundational nature of mathematics as a driving force behind technological change, is giving serious attention to these concerns. (See, in particular, [6, 14, 15, 32, 33, 42, 48, 52].) Although some problems are beyond the scope of college faculty to address directly [20], mathematicians can develop and implement a college mathematics curriculum that meets current needs and gives cogent attention to methodologies for teaching mathematics.

During the 1950s, the Mathematical Association of America (MAA) organized its Committee on the Undergraduate Program in Mathematics (CUPM) to express and shape consensus in the mathematical community concerning the undergraduate curriculum. CUPM recommendations about the undergraduate mathematics major appeared in 1965 [17], 1972 [12], and 1981 [54]. The CUPM Subcommittee on the Undergraduate Major (SUM) was created in 1987 to update the 1981 publication. After a careful review of the 1981 report and its relationship to previous reports, SUM solicited views from broad sectors of the mathematics community, placed calls in national publications [8, 50], held public hearings, exchanged information with committees having related charges, and encouraged dialogue both at its meetings and by correspondence. The Subcommittee's report was adopted by CUPM at its January 1991 meeting.

The curriculum recommendations that follow present both a philosophy to guide curriculum development and a framework on which to construct departmental requirements for the undergraduate mathematical sciences major. Issues of teaching and advising are natural adjuncts to the discussion. The program structure takes into account the realities of current practice and reflects successful experience. This common framework provides the opportunity for choices among a broad variety of courses while assuring both depth and flexibility. This report is offered in a helpful spirit as a practical tool for those who want to provide strong preparation for students having widely varying goals. By issuing these recommendations at this time of curricular revitalization at all levels in mathematics education, CUPM hopes to help sustain the momentum for reform in undergraduate mathematics and to provide answers for those who want assistance in their efforts to bring about effective change.

Philosophy

Students who complete mathematics majors have often been viewed by industry, government, and academia as being well-prepared for jobs that require problem solving and creative thinking abilities. The philosophical views that underlie the concrete recommendations of this report provide a basis to ensure that this reputation is upheld and enhanced. The first five tenets in this philosophy reflect timely small changes from those in the 1981 CUPM report [54]; the added four reflect new realities and national issues.

I. **Attitudes and skills.** The mathematics curriculum should have as a primary goal developing the attitudes of mind and analytical skills required for efficient use, appreciation, and understanding of mathematics. It should also focus on the student's ability to function as an independent mathematical learner. The development of rigorous mathematical reasoning and of abstraction from the particular to the general are two themes that should run throughout the curriculum.

II. **Program level.** The mathematical sciences curriculum should be designed around the abilities and academic needs of the typical mathematical sciences student while making supplementary material or courses available to attract and challenge particularly talented students.

III. **Interaction.** Since active participation is essential to learning mathematics, instruction in mathematics should be an interactive process in which students participate in the development of new concepts, questions, and answers. Students should be asked to explain their ideas both by writing and by speaking, and should be given experience working on team projects. In consequence, curriculum planners must act to assure appropriate sizes of various classes. Moreover, as new information about learning styles among mathematics students emerges, care should be taken to respond by suitably altering teaching styles.

IV. **Applications and theory.** Applications should be used wherever appropriate to motivate and illustrate material in abstract and applied courses. The development of most topics should involve an interplay of substantive applications, mathematical problem solving, and theory. Theory should be seen as useful and enlightening for all courses in the mathematical sciences, regardless of whether they are viewed as pure mathematics or applied mathematics.

V. **Recruiting.** First courses in a subject should be designed to appeal to as broad an audience as is academically reasonable so that students entering college will be attracted to the discipline by the vitality of subjects at the calculus level. Students in such courses should be led to sense the intellectual vitality of the mathematical approach to problems and of the construction of theories that resolve those problems.

VI. **Concentrations.** A system of tracks within the mathematical sciences major is an appropriate response to the diverse interests of students and to the potential opportunities for undergraduate majors. However, the undergraduate mathematics program, and the courses in it, should be designed in such a way that students are not forced to make choices about their post-baccalaureate lives too early in their undergraduate careers. In particular, all tracks should incorporate the major intellectual components of the discipline—including the interplay of theory and applications, study in depth, and the construction of general theories and proofs—so that bachelor's graduates will retain maximum flexibility in pursuing diverse opportunities for employment or further study.

VII. **Technology.** Computers have transformed the world in which today's undergraduates live and work. Recent advances have made it possible for computers and graphing calculators to play an important and expanding role in the teaching and learning of mathematics, just as they do in mathematical research. Curriculum planners should take advantage of modern computing technology. Beyond being competent in a high-level programming language, mathematical sciences majors should be familiar with symbolic manipulation software. In applied courses such as statistics, numerical analysis, and operations research, they should be exposed to standard software packages used by practitioners.

VIII. **Transitions.** The nurturing of the next generation of doctoral mathematical scientists is a special shared responsibility of all undergraduate and graduate teachers and advisors. College and university faculties should work together to ease the transition for students who move from mathematical sciences majors to the more theoretical focus of study for the master's or doctoral degrees. Professional society channels should be used as a resource both for communications and for recommendations concerning graduate study.

IX. **Advising.** Advising of mathematics majors throughout their undergraduate years is an important responsibility. Careful and sustained individualized advising is a crucial aspect of the student's mathematics program. Creating specialized concentrations within a department's major imposes added advising responsibilities.

The Program

This section describes a general intellectual and support framework for the mathematical sciences major of the 1990s which is designed to implement the tenets of the philosophy enumerated above. The framework is shaped by a curricular *structure* which has fixed *components* but allows considerable latitude in the choice of specific courses. Combined with the creation of specialized curriculum concentrations or *tracks* within the major, this structure provides both flexibility and utility. It does, however, impose a serious advising responsibility on the mathematics faculty. Moreover, since the effectiveness of any curriculum depends strongly on the quality of teaching, mathematicians will need to become aware of research, innovations, and recommendations concerning undergraduate teaching. (See, for example, *Undergraduate Mathematics Education Trends* and many of the references cited below, including [6, 7, 16, 19, 29, 32, 33, 39, 41, 42, 43, 45, 47, 49, 51, 57, 58].)

Advising

Unlike an earlier, simpler day when all mathematics majors took essentially the same sequence of courses with only a few electives in the senior year, the typical undergraduate mathematical sciences department today requires students to make substantial curricular choices. As a result, departments have advising responsibilities of a new order of magnitude. Students need departmental advice as soon as they show interest in (or potential for) a mathematics major. Advisors should carefully monitor each advisee's academic progress and changing goals, and together they should explore the many intellectual and career options available to mathematics majors. Career information (e.g., [28]) that informs students of the wide variety of career options open to them is important. If a "minor" in another discipline is a degree requirement or option, then achieving the best choice of courses for a student may necessitate coordination between the major advisor and faculty in another department.

Advisors should pay particular attention to the need to retain capable undergraduates in the mathematical sciences pipeline, with special emphasis on the needs of under-represented groups [23, 34, 37, 39, 52, 53]. When a department offers a choice of several mathematical tracks within the major, advisors have the added responsibility of providing students with complete information even when students do not ask many questions. Track systems may lead students to make life-time choices with only minimal knowledge of the ramifications; therefore, departments utilizing these systems for their majors must assure careful and timely information. One requisite of an individualized approach to advising is that each advisor be assigned a reasonable number of advisees. (See also the chapter on advising in [42].)

Structural Components

This section describes a framework which is a useful tool for constructing course requirements for a department's mathematical sciences major. This structure involves both specific courses (e.g., "linear algebra") and more general experiences (e.g., "sequential learning") derived through those courses.

Seven components form the structure for a mathematical sciences major:

    A. Calculus (with Differential Equations)
    B. Linear Algebra
    C. Probability and Statistics
    D. Proof-Based Courses
    E. An In-Depth Experience
    F. Applications and Connections
    G. Track Courses, Departmental Requirements, and Electives

The first six components will normally require nine or ten courses, at least seven of which would be taught by the major department. (By a course, we mean a three or four semester-hour course.) Local curricular requirements or the desire to achieve strong preparation may necessitate further specificity in the major. Choices within components should be based on close consultation between student and advisor. To facilitate coherent choices, a department may organize a system of specialized concentrations within the major. Additional courses appropriate to particular tracks are part of the last of the structural components.

Under normal circumstances, every track of a department's mathematical sciences major should require courses fitting each of these seven components. Exceptions would be warranted in those situations where a track offered by a mathematical sciences department conforms to the curricular recommendations of another professional society—for example, a computer science track conforming to current national guidelines (presently [55]) or a statistics track following American Statistical Association (ASA) recommendations. Departments should name such special tracks appropriately, and should inform students of the sources of the curricular recommendations that are the basis for any special tracks.

A. Calculus (with Differential Equations)

All mathematical science major programs should include the content of three semesters of calculus. This requirement is central to the major. There is much interest in the mathematical community in reworking calculus courses originating during the late 1980s [16, 46] and continuing into the 1990s [20, 56]. Several professional societies and funding agencies have focused their attention on calculus, so the resulting prototypes should receive consideration during curriculum planning. Since more client disciplines require calculus than any other course in the major, serious attention to concerns of those disciplines is necessary and appropriate. For example, the most recent recommendations for computer science students include a minimum of two calculus courses which are to include series and differential equations. These recommendations, which were in preparation during the period 1988–1990, also place heavy emphasis on theory, definition, axioms, theorems, and proof as a "process ...present in ...introductory mathematics courses—calculus or discrete mathematics" [55].

Calculus courses show wide variation in arrangement of topics, levels of rigor, and methods of presentation. Central topics from calculus normally include discussion of limits, derivatives, integrals, the Fundamental Theorem of Calculus, transcendental functions, elementary differential equations, Taylor's theorem, series, curves and surfaces, partial differentiation, multiple integration, vector analysis including Green's, Stokes', and the divergence theorems, and experience with approximations, numerical methods, and challenging computations. Better-prepared students may be given much more, and the level of rigor and sophistication will vary substantially from course to course.

This requirement will typically be satisfied by the equivalent of a sequence of three mathematics courses.

B. Linear Algebra

Every mathematical science major should include a course in linear algebra. This requirement serves three distinct purposes. First, it introduces students to a part of mathematics which is the foundation of an ever-widening family of significant applications. Second, it provides a geometrically based preparation for many upper-level courses in algebra and analysis. Third, it serves as a needed bridge between lower- and upper-division mathematics courses. Since first-year calculus is usually taught as a service course with little emphasis on theory, many mathematicians see a balancing emphasis on mathematical structure with an introduction to theorems and proofs as an essential and natural component of linear algebra. (When linear algebra is an integral part of the lower-division component of the major, close consultation between the two- and four-year colleges in a given region is advisable.)

The content of the first linear algebra course is currently under discussion in the mathematical community. The traditional vector-space-oriented linear algebra course begins with matrix calculations, develops vector space ideas starting with R^2, moves to R^n, and then to the study of abstract vector spaces and their linear transformations; it culminates in a brief treatment of eigenvalues and eigenvectors. A second type of course is matrix-based: it emphasizes more applications and computational topics such as LU decomposition and pseudoinverses, and stresses the pervasive use of linear modelling in quantitative disciplines. A composite approach which is receiving increased attention [5] gives the first linear algebra course a more matrix-based focus while maintaining the same level of mathematical rigor as the traditional vector space approach.

Actual applications of linear algebra are rarely small and neat. As a result, many linear algebra software packages have been developed and are widely used by practitioners. How these packages should be used in the introductory courses is not yet well established; however, there should be active experimentation on the use of these mathematical tools in the first linear algebra course.

This requirement will generally be satisfied with the equivalent of one mathematics course, although the essentials of linear algebra may be included in some calculus sequences.

C. Probability and Statistics

Every mathematical sciences major should include at least one semester of study of probability and statistics at a level which uses a calculus prerequisite. Many elements of the chapter on statistics in the 1981 CUPM report [54] are still timely; see also [11, 29].

Since it is a severe limitation to compress an introduction to probability and an understanding of statistics into one course, care must be taken that it not be a course focusing only on statistical theory. Versions of this course can be developed in which about a month of a suggested fourteen-week syllabus is devoted to probability theory; other courses in the curriculum (e.g., discrete methods, modelling, operations research) can be coordinated to include additional basic probability topics [24, 40].

The major focus of this course should be on data and on the skills and mathematical tools motivated by problems of collecting and analyzing data. Data collection and description are important. It meets practical needs and helps motivate the difficult idea of sampling distribution [4, 29]. Students must be trained to look at data and be aware of pitfalls, to see for themselves the results of repeated random sampling and the variability of data [4, 10, 18, 30, 40, 51].

The probability and statistics course description in the 1981 CUPM report included a computer programming prerequisite and a recommendation that instructors "use library routines or pre-written programs as they wish." Because statistical software packages have become so important in the actual uses of statistics in recent years, any statistics course taught now should use a nationally available software package.

This requirement will generally be satisfied with the equivalent of one mathematics course.

D. Proof-Based Courses

All students in a mathematics or mathematical sciences major should have within their program two terms of proof-based mathematics with a calculus-level prerequisite. One of these courses should be in algebra (or discrete mathematical structures) and one in analysis (continuous mathematics).

Such courses can come in a myriad of attractive packages, and it is not the intent of these recommendations to specify categorically which particular courses would be good selections for the various tracks. However, all students should encounter sustained mathematical discussion that involves the concatenation of definitions and theorems to build a substantial mathematical structure. Indeed, such an intellectual configuration has been one of the central paradigms within mathematics during this century, and all mathematics students should be exposed to it, regardless of where their majors will ultimately lead.

Departments will find that this component in the major structure can be filled in quite different ways. Two natural choices within this category are a mature course in modern algebra and a proof-based course in analysis which could be any of what are often entitled real variables, advanced calculus, introduction to functional analysis, or complex analysis. (Departments should strongly recommend completion of both the traditional algebra and analysis undergraduate sequences for students intending to pursue graduate work in a mathematical science.) Some departments offer a very rigorous introduction to linear algebra, and such a course could be used as the algebra part of this component. Additionally, other courses different from the traditional ones mentioned above could fulfill this component. For example, courses in mathematical logic, computational complexity, linear programming, and graph theory can be taught in ways that would make them appropriate.

This requirement will generally be satisfied with the equivalent of two mathematics courses, although a rigorous course used for the linear algebra component might also satisfy the algebra part of this requirement.

E. An In-Depth Experience

Every mathematical sciences major should pursue in depth at least one mathematical area beyond calculus. The aim of this component in the major structure is that the student participate in a sustained two-course sequence in at least one important area of mathematics. For many students a requirement like this is best met by a two-course sequence with a calculus prerequisite such as the traditional abstract algebra or real analysis sequences. In addition to any second course that builds on one of the proof-based courses outlined above, a variety of other combinations would be useful ways to assure this in-depth mathematical experience for some students. Examples include a two-term sequence in probability and statistics, numerical analysis, or combinatorics and graph theory. The critical condition is that there be a coherent sequence with the second course using and building on the first which in turn has a calculus-level prerequisite.

There are many ways to satisfy this requirement with the equivalent of one mathematics course beyond those required to satisfy the four previous components; some choices (e.g., numerical analysis) may involve two additional mathematics courses.

F. Applications and Connections

All mathematics and mathematical sciences major programs should include two types of courses that lie outside the core of mathematics. Both of these courses should emphasize the applications of mathematics and its connectedness to other disciplines. Although many options can be identified for the two courses in this component to meet individual student goals, most students would be served best by one course of each of the following two types.

A course using computers. For a student with minimum computer experience, the best option would be a computer science course that introduces well-structured programming and algorithmic problem solving in the context of a modern programming language, and includes such topics as an introduction to data structures and the analysis of algorithmic complexity. (The standard Computer Science I course such as described in [55] satisfies this requirement.) For a student entering the major with adequate capability in a high-level programming language, this requirement may be fulfilled with a more advanced computer science course or a quantitative course in the mathematical sciences which requires the use of computers to solve pertinent problems. Examples include a course in operations research, linear or nonlinear programming, control theory, statistics, applied mathematics, or a mathematical modelling course. (Such courses may, depending on the institution's structure, be taught either within or outside of the major department; see also [45, 58].)

A course applying mathematical methods. Typically, this will be a quantitative course outside the mathematical sciences that makes substantial use of the material in one or more of the mathematics courses that are part of the previous components. The intent of this component is to give students a taste of how mathematics is applied in another discipline and is meant to go well beyond the examples typically seen in mathematics courses. This course must have significant mathematical usage and be taught outside the mathematical sciences; examples include calculus-based physics, mathematical economics, and mathematical biology.

This requirement involves two courses normally taught in departments outside the mathematical sciences.

G. Track Courses, Departmental Requirements, and Electives

The major will be completed by courses from three other categories. Usually there will be certain designated courses (or a choice among such courses) that are essential for a particular track; some of these may be taught outside the mathematical sciences and, depending on institutional constraints, may or may not be counted as credit in the major. In addition, there may be institution-wide or departmental requirements such as senior projects, seminars, or other broadening experiences that are offered as courses. There may also be undesignated electives, some number of which may be required to complete the major. This is especially the case when a student skips early courses (perhaps through the Advanced Placement Program) or when there are higher than usual credit requirements for the major. Additional courses of particular relevance to the student's goals should be selected with careful faculty advice.

The number of courses for this component will vary by department and track.

The total number of courses needed to satisfy components **A** through **F** is nine or ten, but one or two of these are not normally taught in a mathematical sciences department. A student who begins with Calculus I will typically satisfy **A** through **F** with eight mathematical science courses, some of which may be equivalent to four semester hours in length. Component **G** will typically include from one to four additional courses, some of which may not be in the major department.

Curriculum Tracks

All tracks (concentrations of courses) in a department's mathematical sciences major should begin with a set of common courses; students are not normally sufficiently informed to choose a track wisely until their junior year. Although tracks give the major a particular focus, it can be argued that general mathematical techniques and reasoning skills are, in fact, the primary asset of mathematics graduates as they begin initial jobs or move to graduate study. It may further be argued that early concentration is not as important as some students and faculty appear to believe.

Many colleges offer several formalized clusters of courses that may be chosen by students majoring in the mathematical sciences; these tracks may be offered in one or several mathematical sciences departments. Typically a department will have two or more designated tracks with some choices of courses within each track. Particular attention must be given to the needs of students who initially choose one track and then decide to change, to students lacking familiarity with the academic and technical worlds who may need extra help and time in making choices, and to retention of students from under-represented groups.

There is no track specifically labeled as preparatory for graduate study. With appropriate choices for the structural components, most tracks could provide sound preparation. (See discussion in the section below entitled Future Graduate Study; also see Philosophy tenets numbered VI, VIII, and IX.)

Appropriate choices of courses for each of the structural components of the major will be departmentally designated or worked out in consultation with faculty advisors, taking into account employment and graduate school constraints. The strong basis provided by structural components **A** through **F** gives the opportunity to design meaningful coherent clusters of track courses in structural component **G**. Departments should adhere with tenacity to the component structure as a basis for tracks within the major except when, as noted above, the track follows recommendations developed by another appropriate professional society. Departments in the mathematical sciences should make their faculty and students aware of applicable guidelines related to the tracks offered their students.

The names of tracks vary considerably from one institution to another. For example, some schools label a track much like the one described below under the title "Computational and Applied Analysis" as "Applied Mathematics" or "Engineering Support." Most of the tracks may be broadly considered as options within applied mathematics. Every applied mathematics track should include some fundamental components of "pure" mathematics just as the pure mathematics student should gain some understanding of applied and computational mathematics.

For some tracks, considerable specificity for components **A** through **F** may be desirable; for others, a large number of courses may be needed in the specialized **G** component. As a consequence, some tracks may be unattractive in a setting where there are institutional limitations on the total number of hours which may be required in the major. Here are some examples of tracks that are commonly found in departments of mathematical sciences:

Actuarial Mathematics. Basic courses in accounting and numerical analysis are desirable, as are specialized courses in actuarial or risk theory and life contingencies. Strong programs include differential equations and courses in linear algebra, advanced calculus, computer science and mathematical economics; these courses easily fit in the framework of components **A** through **F**. Departments offering this track will want to keep abreast of the content of actuarial tests and other information available from the actuarial societies. (See, for example, materials of the Casualty Actuarial Society [9], the Society of Actuaries [46], and the journal entitled *The Student Actuary* [2].)

Applied Statistics. For applied statistics, the structural components **A** through **G** usually include analysis, numerical analysis, and discrete mathematics. A balance of theory, methods, and applications courses in statistics may include, at the upper-division level, topics such as experiment design and analysis, regression analysis, quality control, sampling methods, data analysis and multivariate methods, nonparametric statistics, and elective topics. Utilization of statistical software must be emphasized. Courses from the Management Science track are helpful. Courses in Computer Science such as advanced programming techniques (including elementary data structures), operating systems, and computability are also appropriate.

Computational and Applied Analysis. Within **A** through **F**, students should include courses from differential equations, advanced calculus or real analysis, complex analysis, and numerical analysis. Modelling and simulation courses are also important. Strong work in the physical sciences is advisable. Somewhere in the course structure project or team-oriented assignments should be emphasized.

Computer Analysis. A track labeled Computer Science should follow the recommendations of the professional computer societies [55]. Even if offered in a joint or mathematical sciences department, such majors are computer science majors. In a small department, a mathematical sciences major which includes a number of computer science courses but does not meet fully the requirements of [55] might also include a year of rigorous discrete mathematics and additional courses from operations research, modelling and simulation, abstract algebra or discrete structures, combinatorics, and applications areas.

Management Science. Students desiring preparation in management science should select their mathematical electives from the following courses: advanced statistics, decision mathematics, design and analysis of experiments, applied regression, operations research, modelling and simulation, and mathematical programming. In addition, students should consider taking courses in accounting, economics, business law, and principles of management or marketing.

Operations Research. Beyond calculus and linear algebra, recommendations for this track include advanced calculus, two courses in probability and statistics, linear programming, stochastic models, combinatorics or graph theory, and two courses in computer science (the second a course with a substantial amount of data structures). There should also be at least one course in economics. A track with courses very similar to those for Operations Research may be labeled Discrete Mathematics; the two tracks sometimes exist in the same department. Coding theory is included in Discrete Mathematics, and there is often a heavier emphasis on theoretical computer science.

Pure Mathematics. This option is sometimes labeled Core Mathematics and might also be termed Classical Mathematics. One year each of abstract algebra and real analysis should be included; this requires one course more than the minimum number to satisfy components **B** and **D**. Strong additions are courses in topology or geometry, mathematical logic, complex variables, number theory; some numerical analysis or theoretical computer science is advised.

Scientific Computing. A modern thrust in science and mathematics is the use of computers to cast light on disciplinary issues. This has led to a new subject area within the mathematical sciences in which students study computational and graphical techniques with an eye to their application in other scientific or mathematical subject areas. Within this option, typical mathematics courses after calculus and linear algebra include ordinary, partial and numerical differential equations, optimization, and modelling. Modern programming languages, data structures, parallel and vector processing, computer graphics, computer simulation, and software engineering are important courses from computer science, as is familiarity with a variety of important computational software packages.

Systems Analysis. Students with general interest in systems analysis or information science should choose courses from the Management Science track. In addition, they should consider courses in accounting, application programming, advanced programming techniques, software engineering, databases and file structures, and management information and decision support systems. This track should emphasize project-oriented assignments.

Teaching (Secondary). After the full calculus sequence, real analysis (or advanced calculus) should be included in this track because graduates need this preparation to teach strong calculus courses in the high school curriculum. In addition, courses in Euclidean and non-Euclidean geometry, abstract algebra, modelling, history of mathematics, and number theory are very useful for the prospective secondary school teacher. (More specific information is contained in the COMET and NCTM guidelines [25, 31].)

Completing the Major

The mathematical sciences major should also involve a variety of other types of experiences and activities that are, in some cases, "co-curricular." Most of the following program elements will help mathematical sciences graduates communicate with colleagues, work with groups of people with differing backgrounds, and build their mathematical self-confidence. (See also [2, 3, 19, 33, 39, 41, 43, 48, 49, 53, 57].)

Integrative Experiences

Every student who majors in mathematics should be encouraged to think about the discipline as a whole. The distinctive nature of mathematical activity, its history and place in the intellectual realm, and its usefulness in society at large are matters on which majors should reflect. This aim may be addressed in a variety of different ways by departments and institutions. The form such an activity assumes is highly dependent on departmental faculty and their interests. For example, some colleges require a senior independent project: this could be a modelling project (perhaps with government or industry linkage) in an applied area of the student's interest, a research or scholarly investigation in an area of mathematics, or an independent study that includes writing about some area of mathematics. Other examples include problem-oriented senior seminars, an undergraduate colloquium series, or seminars in which students present accessible journal articles. Most of these "capstone experiences" include reflection, writing, and oral communication of mathematics. These opportunities are sometimes the most rewarding and enlightening mathematical experience a major has, so they should be encouraged and supported in all appropriate circumstances.

Communication and Team Learning

Most mathematical sciences majors will eventually work in situations where success requires that they be able to explain mathematical work to non-mathematical audiences. In these situations the ability to communicate clearly and effectively about mathematics is vital. Undergraduate mathematical sciences programs should prepare students for communicating mathematics, both orally and in writing. Student presentations in mathematics clubs, in special seminars, and in regular classes can provide practice to help develop skills in communication. Similarly helpful are joint class projects requiring team work with results that are to be well-presented in standard English. In appropriate courses, students can be assigned to teams for collaborative investigation of mathematical problems of some complexity. Such problems should encourage verbal interaction among the team members as well as the written expression of mathematical ideas. Communication skills are further honed when upper-division mathematics students assist freshmen and sophomores in learning, or tutor high school students, perhaps in student service programs.

Independent Mathematical Learning

Whether a mathematics major enters a mathematical sciences career immediately after graduation, goes directly to graduate school, or enters another career path, the student will need to function as an independent learner. Imaginative curriculum and teaching can speed the development of this independent mathematical learning. Readily accessible methods include increasing the numbers of student presentations in junior and senior courses, or independent study projects in which students are expected to master a mathematical subject in the strictly elective part of the curriculum. (However, no part of the recommended component courses **A** through **F** should be replaced by an independent study activity.)

Structured Activities

On many campuses, a variety of organized activities are available to strengthen the undergraduate experience. Student success has been shown to be supported by linking out-of-class activities with study [53]. Such supplementary activities must be set up in a way that is sensitive to the need to be inclusive of under-represented groups, and not just be neutrally available, if these students are to benefit.

For many years strong students have participated in mathematical honorary societies. On other campuses, mathematics club-type activities are available and include a broad range of students and interests. More recently, MAA has developed a Student Chapter program which encourages participation by all students with interests in the mathematical sciences. Within college honors programs, specialized mathematics activities may be available. Mathematical contests may provide valuable broadening with the opportunity for team work and independent learning. In addition to the long-established Putnam competition and the newer Mathematical Contest in Modelling [36], local and regional contests at both lower- and upper-division levels hone mathematical skills.

Undergraduate research experiences [43] including summer research programs are valuable both as learning experiences and as identification and recruiting devices to aid retention in the academic setting of talented students. Stipends are often available to students for summer research and internship activities; NSF funds a number of undergraduate research programs at universities, and government and industrial organizations have found that summer internships are advantageous both to the participating students and the sponsoring organizations. Internships during the academic year are less common in mathematics than in engineering, but where available they provide students both funds for future terms in school and valuable work experience. At some institutions, cooperative education experiences are available for interested students.

Additional Considerations

Our recommendations for the undergraduate major in the mathematical sciences arise out of current priorities and practice within the profession. The major depends on curriculum, on teaching practices, and on student experiences that move beyond strictly classroom learning. It must also be based on a realistic assessment of the needs of the profession—particularly for well-qualified and enthusiastic teachers of mathematics at all levels. The following sections outline some related concerns that the mathematical community must keep in mind in conjunction with the undergraduate major.

Secondary Teacher Preparation

Mathematical sciences departments have always played a significant role in the preparation of secondary teachers of mathematics. Courses designed according to the philosophies and practices described in this report will enrich the experiences needed by teachers who will be confronted with changing demands on the curriculum in the schools. The MAA's Committee on the Mathematical Education of Teachers [25] and the National Council of Teachers of Mathematics [31] both recommend that prospective secondary school teachers of mathematics complete a major in mathematics. Thus students who have selected secondary school teaching as a profession will select courses appropriate to this objective while satisfying the structural components of the mathematical science major. Meeting the needs of those students who seek teaching credentials can be facilitated by flexibility in the department's implementation of a component-based major. For example, courses in Euclidean and non-Euclidean geometry, number theory, or the history of mathematics are useful for prospective teachers and are often required for state certification; such courses can easily be included within an appropriate track structure.

Future Graduate Study

Any track in a mathematical sciences major following the component structure recommended in this report can lead to successful graduate study for students showing potential for such study. It is important that early specialization not preclude a strong foundation for future study. The concerns that a reasonable number of undergraduates pursue graduate work, that they form a demographic cross section, and that they be well-informed about ways to make the transition to graduate school smoothly, are discussed in other parts of this report and in [6, 21, 26, 44].

For students who may wish to pursue graduate study, the courses chosen to satisfy components **D** and **E** will be especially critical; an expectation of mathematical maturity and sophistication is justified for these students. Data from a survey [35] of the highest-ranked graduate departments (including some applied mathematics departments) show that—independent of the graduate specialty—72 of 76 reporting departments rated real analysis or advanced calculus as essential or highly recommended preparation for their programs.

Information Needs

To make wise choices, one needs good information about the effects of different options based on past experience. Unfortunately, in the area of undergraduate mathematics, good information is rare. Here are some examples of areas where much better information is urgently needed:

Teaching Methods. Information on effective and efficient teaching methods for the college mathematics classroom is sparse; while there is some information in journals and periodicals and some in the recent helpful MAA volumes (e.g., [7, 42, 56]), too little specific information of this kind is regularly collected and published. There are, however, many recommendations encouraging improvement and even commentary criticizing both teaching methods and the professoriate, not only in the popular papers but also in the professional literature. It is often recommended (e.g., [6, 25, 32, 33, 48]) that teaching methods other than lecturing should be used. To help those charged with curricula reform on campus, practical information is needed on effective teaching styles for various learner groups.

Educational Environment. Continued efforts should be encouraged for professional societies to collect and disseminate information about ways to improve the environment of the mathematical curriculum. Specific needs noted as this report was prepared include requests for information on the nurturing and encouragement of junior faculty, on evaluation of teaching effectiveness, and on advising strategies, particularly regarding job and further study opportunities for mathematics majors.

Major Requirements. There is not much current data about the actual status and requirements of mathematical sciences major programs in this country. The quintennial sample survey data of the Conference Board of the Mathematical Sciences (CBMS) provide partial snapshots, but not in sufficient detail about the major. It would also be helpful to have more information about subsequent experiences of former students.

Existing Tracks. More information needs to be gathered about existing tracks within the mathematical sciences major. Some tracks may impose only a few restrictions on the choice of courses within the structural components but require a number of courses outside the mathematical sciences. Other tracks may impose very specific requirements on the basic structural components but allow a high degree of choice otherwise. Dialogue between the mathematical community and the clients of various tracks would help ensure appropriate coordination of requirements. (For example, discussion with actuaries may help align actuarial mathematics curriculum recommendations with the advice both of those who make the actuarial examinations and of those who hire actuarial trainees.)

Graduate School Preparation and Retention. National data and anecdotal information suggest that many U.S. undergraduates who enter graduate school do not persist to graduate degrees. If that is true, then it should be a high priority of professional societies, reflecting faculty concern, to recommend improvements. Curriculum planners at the undergraduate level and graduate faculty need better information if they are to work together effectively to improve success rates of beginning graduate students. (See [1, 13, 14, 15, 21, 26, 27].)

Statistics. A joint committee of the American Statistical Association (ASA) and the MAA is currently examining a wide range of issues, including the probability and statistics course suitable for mathematics majors and other undergraduate statistics courses taught in (joint) mathematical sciences departments. It would be helpful if this joint committee were to publish guidelines for an undergraduate statistics major or a statistics track within a mathematical science major.

Conclusions

Programs that follow the recommendations of this report will develop mathematical sciences graduates with broad experience and good prospects for using and continuing to learn mathematics. Whether a student intends to teach in a secondary school, work in business or industry, or pursue graduate study, an exciting and flexible program fashioned from the components of this major structure will be appropriate for long-term goals and unanticipated challenges. As with any set of rules intended to apply in diverse situations, flexibility may be warranted in special cases (such as majors within a mathematical sciences department that follow another professional society's guidelines).

The philosophy expressed in these recommendations embodies educational principles that will lead to an enriching educational experience, and the recommended program structure provides a flexible vehicle for fulfilling those principles. One underlying tenet, however, transcends the particular form of curriculum implementation: It is only by requiring substantive achievement of our students that we will be able to produce the sort of quantitatively expert individuals who are going to be the mainstay of the discipline and of society for the next century.

References

[1] Albers, Donald J.; Anderson, Richard D.; and Loftsgaarden, Don O. *Undergraduate Programs in the Mathematical and Computer Sciences: The 1985–1986 Survey.* MAA Notes No. 7. Washington, DC: Mathematical Association of America, 1987.

[2] American Society of Student Actuaries. *The Student Actuary: The Journal for Actuarial Progress.* Downers Grove, IL 60515. Three issues yearly.

[3] Association of American Colleges. *Integrity in the College Curriculum.* Washington, DC: Association of American Colleges, 1985.

[4] Box, George; Hunter, William; and Hunter, J. Stuart. *Statistics for Experiments.* New York: Wiley, 1978.

[5] Carlson, David H.; Johnson, Charles R.; Lay, David C.; Porter, A. Duane. *The Undergraduate Linear Algebra Curriculum.* Report from an NSF-sponsored Conference (To Appear).

[6] Case, Bettye Anne. "How Should Mathematicians Prepare for College Teaching?" *Notices of the American Mathematical Society,* 36:10 (December 1989) 1344–1346.

[7] Case, Bettye Anne (Ed.). *Keys to Improved Instruction by Teaching Assistants and Part-time Instructors.* MAA Notes No. 11. Washington, DC: Mathematical Association of America, 1988.

[8] Case, Bettye Anne. "Undergraduate Majors in the Mathematical Sciences." *Undergraduate Mathematics Education Trends,* 1 (May 1989) 3.

[9] Casualty Actuarial Society. *Syllabus of Examinations.* Thirty-Fifth Edition. New York: Casualty Actuarial Society, 1990.

[10] Cleveland, W.S. *The Elements of Graphing Data.* Belmont, CA: Wadsworth, 1985.

[11] Cobb, G.W. "Introductory Textbooks: A Framework for Evaluation." *Journal of the American Statistical Association,* 82 (1987) 321–339.

[12] Committee on the Undergraduate Program in Mathematics. "Commentary on 'A General Curriculum in Mathematics for Colleges'." Washington, DC: Mathematical Association of America, 1972. (See CUPM *Compendium,* 1973, pp. 33–101.)

[13] Connors, Edward A. "1989 Annual AMS-MAA Survey." *Notices of the American Mathematical Society,* 36:9 (November 1989) 1155–1188.

[14] David, Edward E., Jr. "Renewing U.S. Mathematics: An Agenda to Begin the Second Century." *Notices of the American Mathematical Society,* 35 (October 1988) 1119–1123.

[15] David, Edward E., Jr. *Renewing U.S. Mathematics: A Plan for the 1990s.* Washington, DC: Committee on the Mathematical Sciences, 1990.

[16] Douglas, Ronald G. (Ed.). *Toward a Lean and Lively Calculus.* MAA Notes No. 6. Washington, DC: Mathematical Association of America, 1986.

[17] Duren, W.L., Jr. *A General Curriculum in Mathematics for Colleges.* Washington, DC: Mathematical Association of America, 1965.

[18] Freeman, D.; Pisani, R.; Purves, R. *Statistics,* 2nd ed. New York: W.W. Norton, 1978.

[19] Gopen, George D. and Smith, David A. "What's an Assignment Like You Doing in a Course Like This? Writing to Learn Mathematics." Reprinted in *The College Mathematics Journal,* 21 (1990) 2–19.

[20] Halmos, Paul R. "The Calculus Turmoil." *Focus,* 10:6 (Nov.-Dec. 1990) 1–3.

[21] Jackson, Allyn. "Mathematics Job Market: Are New Ph.D.s Having Trouble Getting Jobs?" *Notices of the American Mathematical Society,* 37:10 (December 1990) 1349–1352.

[22] Karian, Zaven (Ed.). *Report of the CUPM Subcommittee on Symbolic Computation.* Washington, DC: Mathematical Association of America (To Appear).

[23] Kenschaft, Patricia Clark. *Winning Women into Mathematics.* Washington, DC: Mathematical Association of America, 1991.

[24] Larsen, R.J. and Marx, M.L. *An Introduction to Mathematical Statistics and Its Applications,* 2nd ed. Englewood Cliffs, NJ: Prentice-Hall, 1986.

[25] Leitzel, James R.C. (Ed.). *A Call for Change: Recommendations for the Mathematical Preparation of Teachers of Mathematics.* Washington, DC: Mathematical Association of America, 1991.

[26] Lipman, Joseph. *Notices of the American Mathematical Society,* 37:9 (November 1990) 1207.

[27] Madison, Bernard L. and Hart, Therese A. *A Challenge of Numbers: People in the Mathematical Sciences.* Washington, DC: National Academy Press, 1990.

[28] Mathematical Association of America. *Mathematical Scientists at Work: Careers in the Mathematical Sciences.* Washington, DC: Mathematical Association of America, 1991.

[29] Moore, David. *Statistics: Concepts and Controversies,* 3rd ed. New York: W.H. Freeman, 1991.

[30] Moore, David and McCabe, G. *Introduction to the Practice of Statistics.* New York: W.H. Freeman, 1989.

[31] National Council of Teachers of Mathematics. *Professional Standards for Teaching Mathematics.* Reston, VA: National Council of Teachers of Mathematics, 1991.

[32] National Research Council. *Everybody Counts: A Report to the Nation on the Future of Mathematics Education.* Washington, DC: National Academy Press, 1989.

[33] National Research Council. *Moving Beyond Myths: Revitalizing Undergraduate Mathematics.* Washington, DC: Committee on Mathematical Sciences in the Year 2000, National Academy Press, 1991.

[34] National Science Foundation. *Women and Minorities in Science and Engineering.* Washington, DC: National Science Foundation, 1988.

[35] Neidinger, Richard. "Survey on Preparation for Graduate School." *Focus,* 8:4 (September 1988) 4.

[36] "News and Announcements." *Notices of the American Mathematical Society,* 37:7 (September 1990) 879–880.

[37] Oaxaca, Jaime and Reynolds, Ann W. (Eds.). *Changing America: The New Face of Science and Engineering, Final Report.* Task Force on Women, Minorities, and the Handicapped in Science and Technology, January 1990.

[38] Ralston, Anthony (Ed.). *Discrete Mathematics in the First Two Years.* MAA Notes No. 15. Washington DC: Mathematical Association of America, 1989.

[39] Resnick, Lauren B. *Education and Learning to Think.* Committee on Mathematics, Science, and Technology Education, National Research Council. Washington, DC: National Academy Press, 1987.

[40] Rice, J.A. *Mathematical Statistics and Data Analysis.* Belmont, CA: Wadsworth, 1987.

[41] Schoenfeld, Alan H. *Mathematical Problem Solving.* New York: Academic Press, 1985.

[42] Schoenfeld, Alan H. (Ed.). *A Source Book for College Mathematics Teaching.* Washington, DC: Mathematical Association of America, 1990.

[43] Senechal, Lester (Ed.). *Models for Undergraduate Research in Mathematics.* MAA Notes No. 18. Washington, DC: Mathematical Association of America, 1990.

[44] Simon, Barry (Ed.). *Report of the Committee on American Graduate Mathematics Enrollments.* Washington, DC: Conference Board of the Mathematical Sciences, 1987. Summary in *Notices of the American Mathematical Society,* 34 (August 1987) 748–750.

[45] Smith, David A.; Porter, Gerald J.; Leinbach, L. Carl; Wenger, Ronald H. (Eds.). *Computers and Mathematics: The Use of Computers in Undergraduate Instruction.* MAA Notes No. 9. Washington, DC: Mathematical Association of America, 1988.

[46] Society of Actuaries. *1991 Associateship Catalog.* Schaumburg, IL: Society of Actuaries, 1991.

[47] Steen, Lynn Arthur (Ed.). *Calculus for New Century: A Pump, Not a Filter.* MAA Notes No. 8. Washington, DC: Mathematical Association of America, 1988.

[48] Steen, Lynn Arthur (Ed.). *Challenges for College Mathematics: An Agenda for the Next Decade.* Washington, DC: Mathematical Association of America; *Focus,* 10:6 (November–December 1990).

[49] Sterrett, Andrew (Ed.). *Using Writing to Teach Mathematics.* MAA Notes No. 16. Washington, DC: Mathematical Association of America, 1990.

[50] Subcommittee on the Major in the Mathematical Sciences (Bettye Anne Case). "Information Needed to Strengthen Undergraduate Mathematics Programs." *Focus,* 8:4 (September 1988) 6.

[51] Tanner, M. "The Use of Investigations in the Introductory Statistics Course." *The American Statistician,* 39 (1985) 306–310.

[52] Thurston, William P. "Mathematical Education." *Notices of the American Mathematical Society,* 37:7 (September 1990) 844–850.

[53] Treisman, Philip Uri. "A Study of the Mathematics Performance of Black Students at the University of California, Berkeley." In *Mathematicians and Education Reform.* Providence, RI: CBMS Issues in Mathematics Education, Volume 1, 1990, 33–46.

[54] Tucker, Alan. *Recommendations for a General Mathematical Sciences Program.* Washington, DC: Mathematical Association of America, 1981. (Reprinted as first of six chapters of *Reshaping College Mathematics,* MAA Notes No. 13, 1989.)

[55] Tucker, Allen B. *Computing Curricula, 1990.* Report of the Joint Task Force of the Association of Computing Machinery and the IEEE Computer Society, 1990.

[56] Tucker, Thomas (Ed.). *Priming the Calculus Pump: Innovations and Resources.* MAA Notes No. 17. Washington, DC: Mathematical Association of America, 1990.

[57] Wilf, Herbert S. "Self-esteem in Mathematicians." *The College Mathematics Journal,* 21:4 (September 1990) 274–277.

[58] Zorn, Paul. "Computing in Undergraduate Mathematics." *Notices of the American Mathematical Society,* 34 (October 1987) 917–923.